Cover and Title Pages: Nathan Love

www.mheonline.com/readingwonders

B

The **McGraw·Hill** Companies

Education

Copyright © 2014 The McGraw-Hill Companies, Inc.

Send all inquiries to:
McGraw-Hill Education
Two Penn Plaza
New York, New York 10121

ISBN: 978-0-02-119111-6
MHID: 0-02-119111-5

Printed in the United States of America.

5 6 7 8 9 DOW 17 16 15 14 13

McGraw-Hill Reading

Wonders

McGraw Hill Education

Bothell, WA • Chicago, IL • Columbus, OH • New York, NY

CCSS Reading/Language Arts Program

Program Authors

Diane August

Donald R. Bear

Janice A. Dole

Jana Echevarria

Douglas Fisher

David Francis

Vicki Gibson

Jan Hasbrouck

Margaret Kilgo

Jay McTighe

Scott G. Paris

Timothy Shanahan

Josefina V. Tinajero

 Education

Bothell, WA • Chicago, IL • Columbus, OH • New York, NY

Unit 1

Growing and Learning

The Big Idea

How can learning help us grow?. **16**

SOCIAL STUDIES

(tl) Stockdisc/PunchStock; (tr) Stockbyte/PunchStock ; (c) John Hovell; (b) Richard Johnson

Go Digital! Find all lessons online at http://connected.mcgraw-hill.com/

Week 3 · Communities 50

Week 4 · Inventions 66

Week 5 · Landmarks 82

(t) Margaret Lindmark

Unit 2

Figure It Out

The Big Idea

What does it take to solve a problem? **96**

(t) Elwood Smith; (c) Janet Broxon; (b) Tristan Elwell

Unit 3

One of a Kind

The Big Idea

Why are individual qualities important?....**176**

(t) Jago Silver; (b) Peter Ferguson

Go Digital! Find all lessons online at http://connected.mcgraw-hill.com/

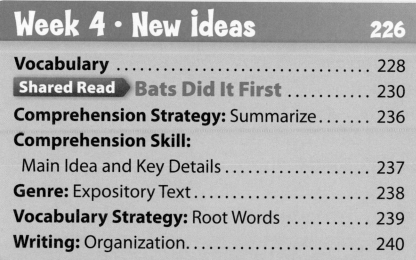
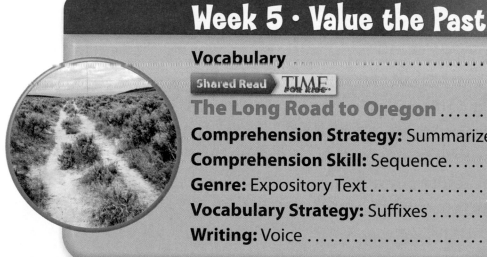

(t) Michael Durham/Minden Pictures; (b) N k Wheeler/Alamy

9

Unit 4

Meet the Challenge

The Big Idea
What are different ways to meet challenges? . .**256**

(t) Michael Moran; (c) Gerardo Suzan ; (b) Marcin Piwowarski

Go Digital! Find all lessons online at http://connected.mcgraw-hill.com/

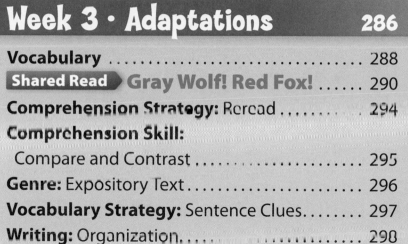

Unit 5

Take Action

The Big Idea

(t) Melissa McGill ; (c) (b) Chris Vallo

Go Digital! Find all lessons online at http://connected.mcgraw-hill.com/

Unit 6
Think It Over

The Big Idea

Go Digital! Find all lessons online at http://connected.mcgraw-hill.com/

Growing and Learning

The More I Know

The more I know,
The more I grow.

It's an easy equation to solve if I try,
Asking and learning help me understand WHY.

The more I know,
The more I WANT to know.

It's an easy equation without a doubt,
Wondering and learning help me figure things out.

— George Samos

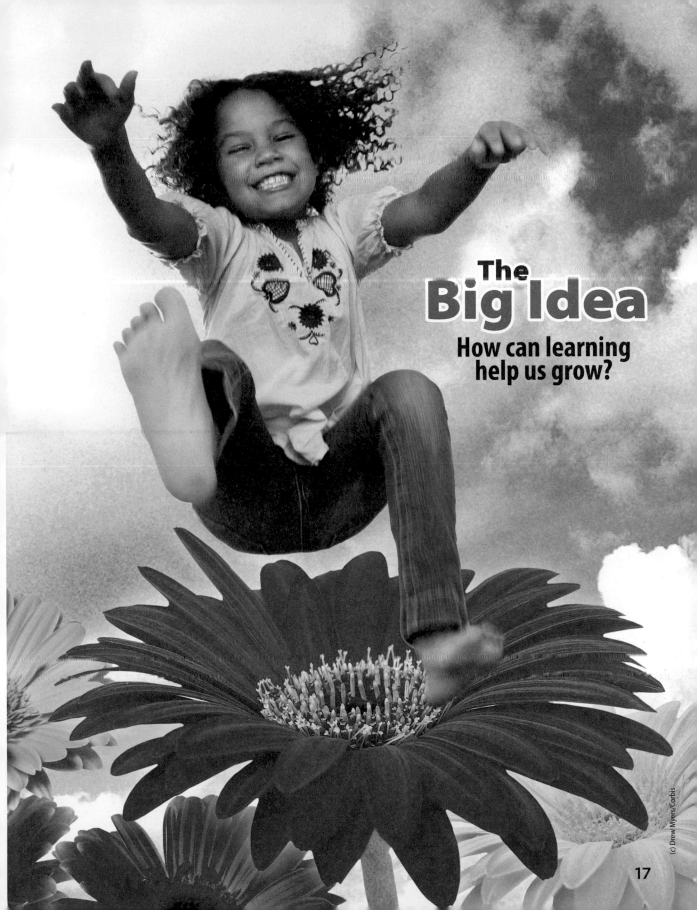

The
Big Idea
How can learning help us grow?

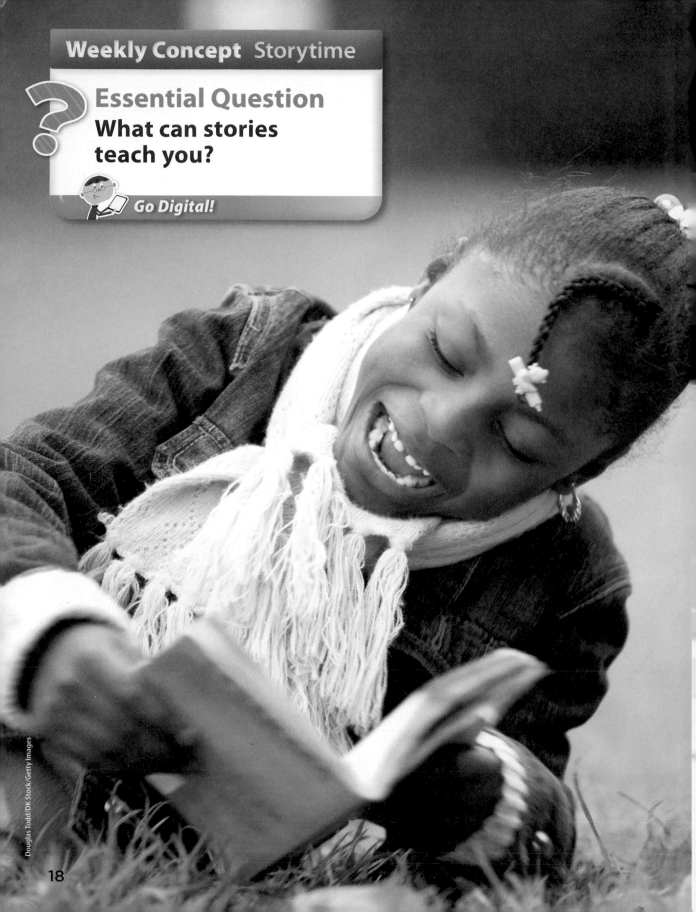

Essential Question
What can stories teach you?

Go Digital!

Read
and Learn

Hi. I'm Katie. Reading is so much fun. I love to read stories for many different reasons.

▶ Stories make me laugh.

▶ They teach me how to do new things.

▶ Stories inspire me to learn about other people.

Talk About It

Write words you have learned about stories. Talk with a partner about other things you can learn from stories.

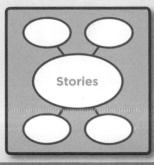

1 of 1

Vocabulary

Use the picture and the sentence to talk with a partner about each word.

ached

My head **ached** and hurt when I had a bad cold.

When has your head ached?

concentrate

Tom can **concentrate** in the quiet library.

Name places where you can concentrate.

discovery

Matt and June made a fun **discovery** at the beach.

What fun discovery have you made?

educated

Jan went to school to be **educated**.

Why does someone become educated?

effort

Learning to ski takes lots of hard work and **effort**.

Name something else that takes effort.

improved

Sam's soccer skills **improved** with practice.

Tell how you improved at doing something.

inspired

My grandmother **inspired** me to learn to knit.

Who has inspired you to try something new?

satisfied

I am happy and **satisfied** with the good job I did.

What word means almost the same as satisfied?

COLLABORATE

Your Turn

Pick three words. Then write three questions for your partner to answer.

Go Digital! *Use the online visual glossary*

BRUNO'S NEW HOME

Essential Question
What can stories teach you?

Read how one story taught a
bear an important lesson.

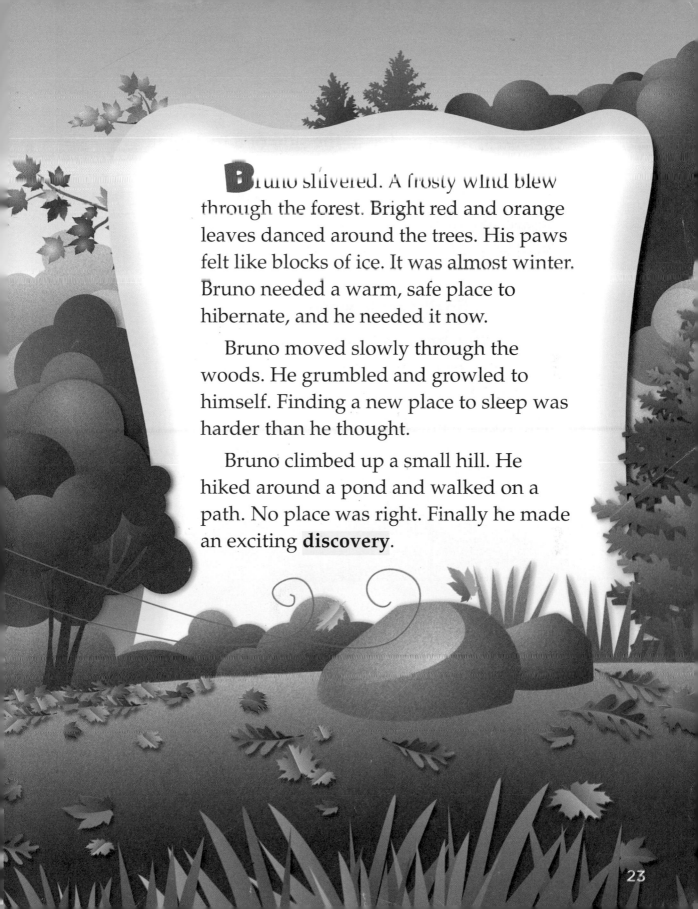

Bruno shivered. A frosty wind blew through the forest. Bright red and orange leaves danced around the trees. His paws felt like blocks of ice. It was almost winter. Bruno needed a warm, safe place to hibernate, and he needed it now.

Bruno moved slowly through the woods. He grumbled and growled to himself. Finding a new place to sleep was harder than he thought.

Bruno climbed up a small hill. He hiked around a pond and walked on a path. No place was right. Finally he made an exciting **discovery**.

Bruno spotted a cave in the side of a rocky hill. It was perfect except for one thing. He couldn't fit through the opening. It was blocked with a pile of dirt and tangled roots.

Bruno sat down to think. "I'll just dig out some of this dirt and make the entrance bigger," he thought. "Then I can finally get to sleep."

Bruno dug and dug with his huge paws, but the dirt was packed down hard. It took a lot of hard work and **effort**. He pulled and pulled at the tree roots.

The roots were strong and too tough to rip out. Bruno stopped to rest. His paws **ached**. They were red and sore. Suddenly he heard a loud sound. "Crack!"

Bruno turned quickly and saw a small squirrel eating a nut. He stopped eating and smiled at Bruno.

"You look like you need help," said the squirrel.

Bruno sighed. "I have been trying to fit into this cave, but it's hopeless. I've been digging and digging, but I haven't **improved** the opening at all."

"I'm Jack, and I can help," said the squirrel.

"But you are too small," said Bruno.

Jack told Bruno to sit down and rest. Bruno sat and yawned as Jack scampered away. A few minutes later, he came back.

"What you need is a book," said Jack. "Reading can help you become **educated**. Books are filled with knowledge." He handed Bruno a thick red book.

Bruno moved to a bright, sunny spot. He put on his glasses and tried to **concentrate** on the story. He paid careful attention to the plot.

The story was about a lion and a mouse. The lion thought the mouse was too small to help him. One day the lion got caught in a net. The mouse chewed the net's ropes and helped the lion escape.

John Hovell

"Well, the lion in this story learned an important lesson," said Bruno. "I think I did, too."

The story **inspired** Bruno. The mouse had sharp teeth, and so did Jack. Jack could help.

The new friends made a fine team. Jack chewed through the thick roots and Bruno dug out the dirt. They worked together all afternoon. Finally, Bruno could fit through the opening.

"Are you **satisfied** and happy with your cozy new home?" asked Jack.

"I sure am!" said Bruno. "And I learned something, too. Good friends come in small packages."

Make Connections

Talk about the story of the lion and the mouse. How did it help Bruno solve his problem? **ESSENTIAL QUESTION**

Discuss how you and your friends help one another. **TEXT TO SELF**

Visualize

Look for colorful words as you read "Bruno's New Home." Use these words to visualize, or form pictures, in your mind.

Find Text Evidence

Why does Bruno need to find a new home? Use the details in the first paragraph on page 23.

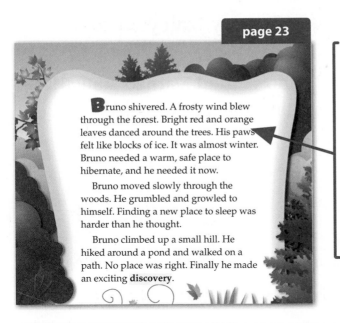

page 23

Bruno shivered. A frosty wind blew through the forest. Bright red and orange leaves danced around the trees. His paws felt like blocks of ice. It was almost winter. Bruno needed a warm, safe place to hibernate, and he needed it now.

Bruno moved slowly through the woods. He grumbled and growled to himself. Finding a new place to sleep was harder than he thought.

Bruno climbed up a small hill. He hiked around a pond and walked on a path. No place was right. Finally he made an exciting **discovery**.

Bruno shivered. A frosty wind blew through the forest. His paws felt like blocks of ice. From these details, I can figure out how cold Bruno feels. He needs to find a warm home so he can hibernate.

COLLABORATE

Your Turn

How does Bruno make the cave's entrance bigger? Reread the story. Visualize what happens. Then answer the question.

Character

A character's actions and feelings make the events in a story happen. Traits are the special ways the character behaves.

 Find Text Evidence

Bruno is the main character in the story. I will reread page 23 to find out what he wants and how he feels.

Character: Bruno	
Wants or Needs	**Feelings**
Bruno needs a new home for the winter.	Bruno is cold, tired, and grumpy.
Actions	**Traits**

Your Turn COLLABORATE

Reread "Bruno's New Home." Think about what Bruno does next. What are his actions? What are his traits? Visualize and fill in the graphic organizer.

Go Digital!
Use the interactive graphic organizer

Fantasy

The story "Bruno's New Home" is a fantasy. A **fantasy**:
- Has characters, settings, or events that do not exist in real life
- Has illustrations that help tell the story
- Teaches a lesson

 Find Text Evidence

I can tell that "Bruno's New Home" is a fantasy. The characters are animals that talk and read. The story teaches a lesson. Illustrations show the story could not happen in real life.

page 25

Bruno turned quickly and saw a small squirrel eating a nut. He stopped eating and smiled at Bruno.

"You look like you need help," said the squirrel.

Bruno sighed. "I have been trying to fit into this cave, but it's hopeless. I've been digging and digging, but I haven't **improved** the opening at all."

"I'm Jack, and I can help," said the squirrel.

"But you are too small," said Bruno.

Jack told Bruno to sit down and rest. Bruno sat and yawned as Jack scampered away. A few minutes later, he came back.

25

Illustrations Illustrations give more information. They show the characters doing things that are not real.

 COLLABORATE

Your Turn

Describe two examples from "Bruno's New Home" that help you know it is a fantasy. Tell your partner.

Synonyms

Synonyms are words that have the same meaning. In "Bruno's New Home," you may read a word you don't know. Use its synonym to figure out what it means.

 Find Text Evidence

On page 23, I'm not sure what grumbled *means. I see the word* growled *in the same sentence. I know that* growled *means "to make a deep, rumbling sound." I think* grumbled *and* growled *are synonyms. They have almost the same meanings. Now I know what* grumbled *means.*

He grumbled and growled to himself.

Your Turn

Find synonyms for these words from "Bruno's New Home."

bright, *page 26*

satisfied, *page 27*

Talk about what each word means with a partner.

John Hovell

Readers to...

Writers use interesting details in their writing to share ideas. They focus on a single moment or event. Reread this passage from "Bruno's New Home."

Focus on an Event

Find one event in the story. What details did the author use to tell about the event?

Expert Model

Bruno dug and dug with his huge paws, but the dirt was packed down hard. It took a lot of hard work and effort. He pulled and pulled at the tree roots.

The roots were strong and too tough to rip out. Bruno stopped to rest. His paws ached. They were red and sore.

John Hovell

Writers

Katie wrote a story about her cat. Read Katie's revisions.

Editing Marks

≡ Make a capital letter.

/ Make a small letter.

⊙ Add a period.

⌃ Add.

⌇ Take out.

Grammar Handbook

Sentence Types
See page 474.

Student Model

☀A Sunny Day

Mama Cat sat on the walk.

The ⌃hot sun was shining⊙ She stretched

and smiled⌇ I saw her four kittens

playing in the cool grass. Then a

dark cloud blew across the sun. I

felt ⌃a big, wet drop of rain on my head. It started to

rain. It was time to go inside⊙

By Katie M.

Your Turn

✔ Identify an event.
✔ Find the details.
✔ Tell how revisions improved Katie's writing.

Go Digital!
Write online in Writer's Workspace

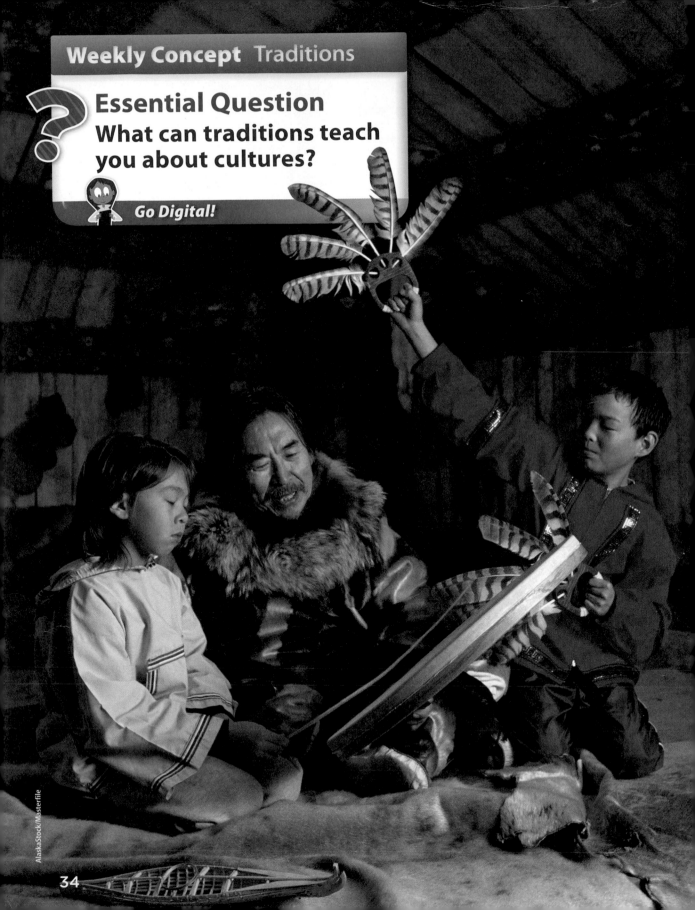

Essential Question
What can traditions teach
you about cultures?

Go Digital!

AlaskaStock/Masterfile

34

SHARING TRADITIONS

My family and I live in Alaska. Today my grandpa is teaching us how to drum and use traditional Yupik dance fans.

▶ Traditions are passed down in my family.

▶ Traditions help me learn about my culture and customs.

▶ My family's traditions make me proud.

Talk About It

Write words you have learned about traditions. Then talk with a partner about how your family shares traditions.

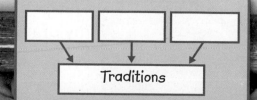

Traditions

Vocabulary

Use the picture and the sentence to talk with a partner about each word.

celebrate

Jim and his friends like to **celebrate** the Fourth of July together.

How do you like to celebrate?

courage

Firefighters show bravery and **courage**.

What word means the same as courage?

disappointment

Jason felt **disappointment** when it rained.

How would you show disappointment?

precious

This **precious** necklace is special to my grandmother.

What does the word precious mean?

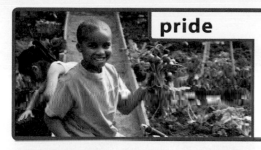

pride

I take **pride** in the vegetables I grow in my garden.

Name one time you felt pride in something you did.

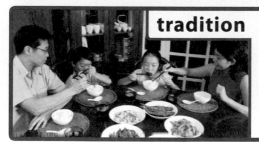

remind

Mom will **remind** me to clean my room.

When do you need someone to remind you to do something?

symbols

The eagle and the flag are **symbols** of our country.

Name some symbols you know.

tradition

Our Sunday **tradition** is eating dinner together.

Name a tradition that people share.

COLLABORATE

Your Turn

Pick three words. Then write three questions for your partner to answer.

Go Digital! *Use the online visual glossary*

(t) Jupiter Images/FoodPix/Getty Images; (tc) Corbis Yellow/Corbis; (bc) Steve Allen/Photographer's Choice/Getty Images; (b) Asia Images Group/Getty Images

The Dream Catcher

? Essential Question

What can traditions teach you about cultures?

Read how Peter learns about his culture.

Richard Johnson

Peter walked home from school. Salty tears ran down his cheeks, and his stomach hurt. He didn't know what to do. Grandmother was waiting for him on the front porch.

"What's wrong, Biyen?" said Peter's grandmother. Biyen was the Ojibwe name for Peter. He called her Nokomis, which means grandmother.

Peter looked up. "I have to give a presentation where I talk about a family **tradition**. I know we have lots of beliefs and customs. Can you **remind** me of one?"

Nokomis smiled and nodded her head.

"Come with me," she said.

Peter followed Nokomis. She went to a closet and stretched to reach the top shelf. She pulled out a small box and blew away the dust. She handed it to Peter.

"Open it," she said.

Peter opened the box. He spotted a wooden hoop inside. It was in the shape of a circle. String was woven and twisted around the hoop. It looked like a spider web. A black bead sat near the center. Feathers hung from the bottom.

Peter wiped away his tears and smiled.

Richard Johnson

"This is a dream catcher," said Nokomis. "Our people have made these for many years. Circles are **symbols** of unity and strength. Let's hang it over your bed tonight. It will catch your bad dreams in the web, and your good dreams will fall through the center. Maybe it will give you **courage** to do your presentation."

"Can I take this one to school?" asked Peter.

"No, Biyen," said Nokomis. "This dream catcher is **precious**. I got it when I was your age, and it means a lot to me."

Peter felt **disappointment** because he wanted to share the dream catcher with his class.

"We could make you one," said Nokomis.

"I'd like that!" cried Peter.

Nokomis and Peter worked together and made a dream catcher. That night, as he gazed and looked at the dream catcher over his bed, he made a plan.

The next morning he told Nokomis his plan. "I'm going to show my class how to make a dream catcher," he said.

Richard Johnson

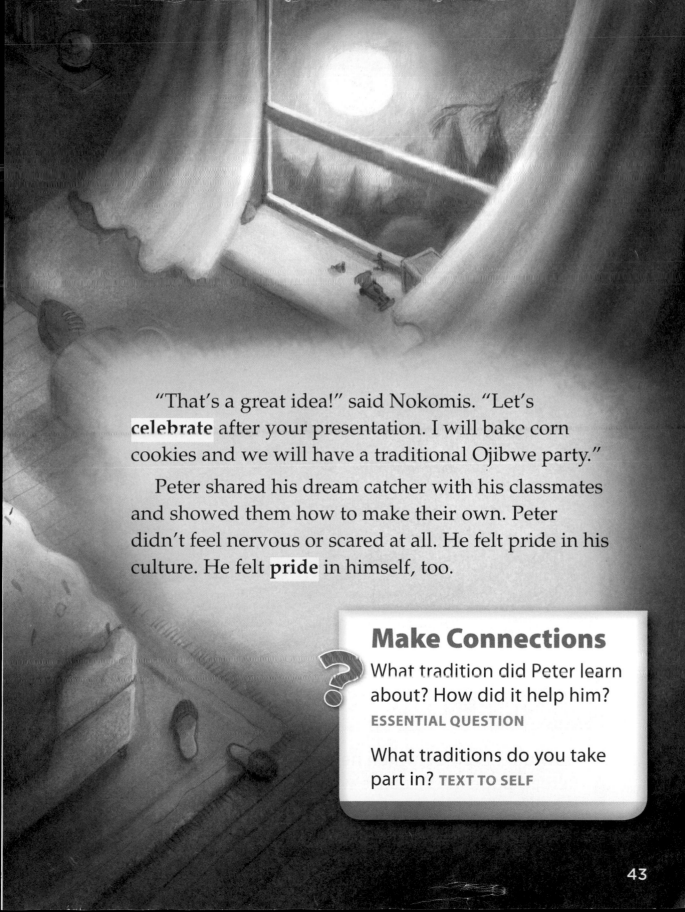

"That's a great idea!" said Nokomis. "Let's **celebrate** after your presentation. I will bake corn cookies and we will have a traditional Ojibwe party."

Peter shared his dream catcher with his classmates and showed them how to make their own. Peter didn't feel nervous or scared at all. He felt pride in his culture. He felt **pride** in himself, too.

Make Connections

What tradition did Peter learn about? How did it help him? **ESSENTIAL QUESTION**

What traditions do you take part in? **TEXT TO SELF**

Visualize

Use details to help you visualize the characters and their actions in "The Dream Catcher." Form pictures in your mind as you read.

Find Text Evidence

How does Peter feel at the beginning of the story? Use the details in the first paragraph on page 39.

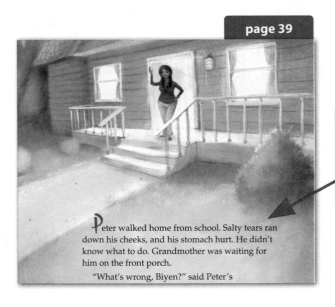

page 39

Peter walked home from school. Salty tears ran down his cheeks, and his stomach hurt. He didn't know what to do. Grandmother was waiting for him on the front porch.

"What's wrong, Biyen?" said Peter's

I can visualize how Peter feels at the beginning of the story. He walked home. Salty tears ran down his cheeks. His stomach hurt. These details help me figure out that Peter feels sad and upset.

Your Turn

COLLABORATE

How did Peter feel when he first saw the dream catcher? Reread page 40. Visualize what happened. Then answer the question.

Richard Johnson

Sequence

A character's actions make up the plot, or events, in a story. Plot events are told in the order they happen. A plot always has a beginning, middle, and end.

 Find Text Evidence

At the beginning of the story, I read to see what the characters say and do. Then I read on to see what happens in the middle of the story.

Character
Peter and Nokomis

Setting
Nokomis's house

The events are in sequence, or time order.

Beginning
Peter is crying. Nokomis asks him what is wrong. He says he has to give a presentation at school.

Middle
Nokomis shows Peter a dream catcher. Then they make one.

End

Your Turn COLLABORATE

Reread pages 42 and 43. What happens at the end of the story? List the events in order in your graphic organizer.

Go Digital!
Use the interactive graphic organizer

Realistic Fiction

"The Dream Catcher" is realistic fiction. **Realistic fiction**:

- Is a made-up story that could really happen
- Has a beginning, middle, and end
- Has illustrations and dialogue

 **Find Text Evidence**

"The Dream Catcher" is realistic fiction. I know because the events could really happen. It also has realistic illustrations and dialogue.

page 39

Peter walked home from school. Salty tears ran down his cheeks, and his stomach hurt. He didn't know what to do. Grandmother was waiting for him on the front porch.

"What's wrong, Biyen?" said Peter's grandmother. Biyen was the Ojibwe name for Peter. He called her Nokomis, which means grandmother.

Peter looked up. "I have to give a presentation where I talk about a family **tradition**. I know we have lots of beliefs and customs. Can you **remind** me of one?"

Nokomis smiled and nodded her head.

"Come with me," she said.

39

Illustrations Illustrations give more information or details about characters and setting.

Dialogue Dialogue is what the characters say to each other.

Your Turn

COLLABORATE

Find two things in the story that could happen in real life. Tell your partner why "The Dream Catcher" is realistic fiction.

Context Clues

If you come across a word you don't know, use context clues. Look at other words in the same sentence. They can help you figure out the word's meaning.

 Find Text Evidence

I read this sentence on page 39. I'm not sure I know what the word presentation *means. I see the words* talk about. *This clue helps me figure out what* presentation *means. A* presentation *is a talk or speech.*

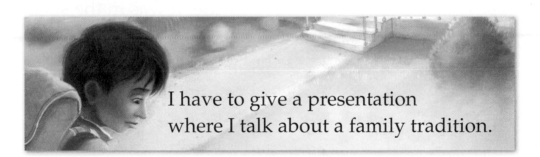

I have to give a presentation where I talk about a family tradition.

Your Turn

COLLABORATE

Find context clues. Figure out the meanings of these words.

woven, *page 40*

gazed, *page 42*

Talk about which nearby words helped you figure out their meanings.

Richard Johnson

Readers to...

Writers use lively words and interesting details to help readers see and feel the events in a story. Reread this passage from "The Dream Catcher."

Details

Find details that describe an event. What words did the author use to tell about the event?

Expert Model

Peter opened the box. He spotted a wooden hoop inside. It was in the shape of a circle. String was woven and twisted around the hoop. It looked like a spider web. A black bead sat near the center. Feathers hung from the bottom.

Peter wiped away his tears and smiled.

Richard Johnson

Writers

Ada wrote a story about a family tradition. Read her revisions.

Editing Marks

⚌ Make a capital letter.

／ Make a small letter.

⊙ Add a period.

∧ Add,

＿ Take out.

Grammar Handbook

Commands and Exclamations
See page 474.

Student Model

♪ My Family Tradition ♪

Every year in May, my family

goes to a big party on our street.

It is so much fun? My grandmother

gives me beads to wear. First we

traditional
eat lots of ∧ food. Then we sing and

dance ⊙ I can't wait! Come ♪

to the party with me.

By Ada H.

Your Turn

COLLABORATE

- ✔ Identify words that describe.
- ✔ Identify a command.
- ✔ Tell how revisions improved the writing.

Go Digital!
Write online in Writer's Workspace

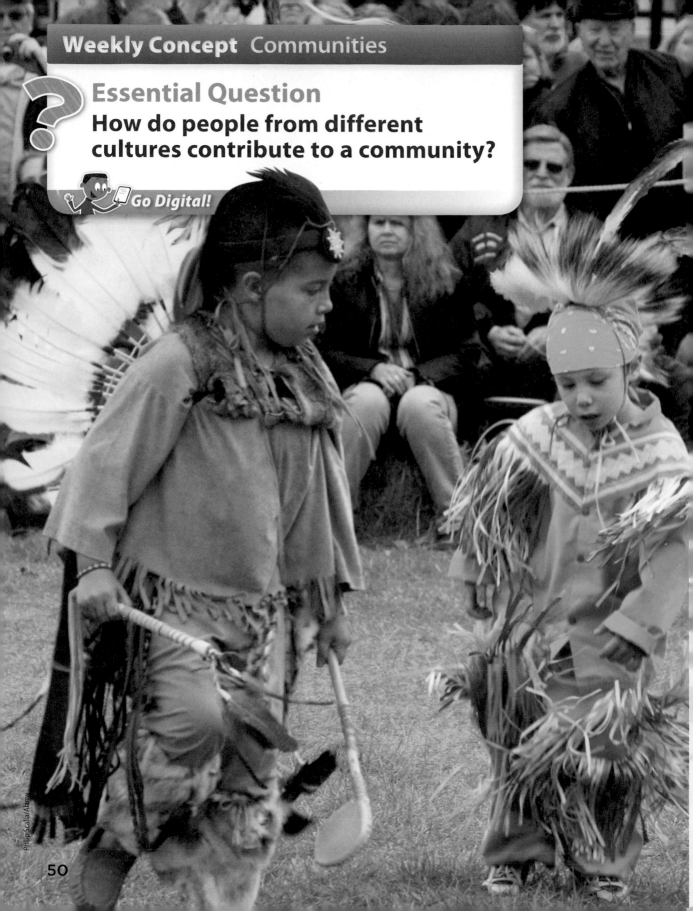

Essential Question
How do people from different cultures contribute to a community?

Go Digital!

MAKING A CONTRIBUTION

David is sharing his culture with the people in his community.

► Learning about different cultures is important.

► Different cultures make a community more interesting.

► Communities grow when people share their cultures.

Talk About It

Write words you have learned about cultures and communities. Talk about what you can learn from other people.

Vocabulary

Use the picture and the sentence to talk with a partner about each word.

admires

My family **admires** my good test grades.

What do you admire about a friend?

classmate

Don and his **classmate** Maria always eat lunch together.

What things do you do with a classmate?

community

Many people in my **community** work together.

What do you like about your community?

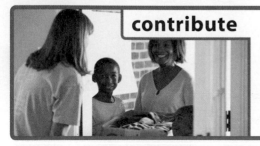

contribute

Mom will **contribute** clothes to people who can use them.

What could you contribute to a bake sale?

practicing

Kyle has been **practicing** and now he can play lots of songs.

What skill can you improve by practicing?

pronounce

Cindy can **pronounce** her name in another language.

How do you pronounce your last name?

scared

Our dog hides during storms because he is **scared**.

What do you do when you feel scared?

tumbled

The ripe tomatoes **tumbled** out of the big basket onto the ground.

What does tumbled mean?

Your Turn

COLLABORATE

Pick three words. Then write three questions for your partner to answer.

Go Digital!   *Use the online visual glossary*

Room to Grow

Our new home in Portland

(flowers) Japack/amanaimagesRF/Corbis; (bkgd) Wetzel and Company; (c) Margaret Lindmark

Essential Question

How do people from different cultures contribute to a community?

Read how one family helps their community grow.

Spring in the City

My name is Kiku Sato. Last spring, my family and I moved from the country to the big city.

Our new home in Portland had no yard. There wasn't even a tiny plot of land. So Mama made an indoor garden. First she and Papa planted seeds in pots. Then they hung them from hooks. Next they crammed plants onto shelves. Green vines **tumbled** over desks. Soon our house had plants everywhere.

At first I was **scared** to start school. I was afraid no one would be my friend. But I soon met a **classmate**. Jill Hernandez and I were **practicing** reading aloud one day. She helped me say her last name, and I helped her **pronounce** mine. The next day we were best friends. Jill spent lots of time at my house.

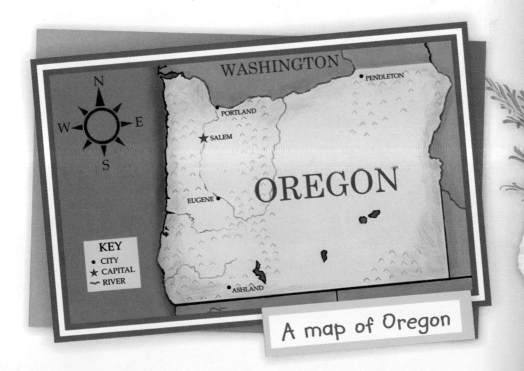

A map of Oregon

An idea for a garden

One afternoon, Jill and her mother came to visit Mama and Papa and me. First they saw our beautiful potted plants. Jill's mother said, "Jill **admires** your indoor garden. She has told me so much about it."

We all sat down while Mama served tea. First she put green tea into the tea bowl. Then she added hot water and stirred. She handed the bowl to Jill's mother and bowed.

Mama's special tea bowls

Grandmother in Japan

"My mother taught me how to make tea," said Mama. "She also taught me how to plant a traditional Japanese garden. I learned to make the most of a small, compact space."

All of a sudden, Jill's mother smiled. "Can you help us with a project?" she asked. "Our **community** wants to plant a garden. Our plot is very small. There is so much we want to grow."

Papa looked at Mama, and they both bowed.

"Yes," they said.

A Garden Grows

First we had a meeting with the community. Everyone agreed to **contribute**. Some people brought seeds, tools, and dirt. Then the next day we met and started our garden.

Papa built long, open boxes. Next, we filled them with dirt. The tallest box went close to the back wall. The boxes got shorter and shorter. The shortest box was in the front. "All the plants will get sunlight without making shade for the others," Mama said.

Papa *building boxes*

Jill and I *planting seeds*

Then, we used round, flat stones to make a rock path. Papa said that in Japan, stones are an important part of a garden. Finally, we planted the seeds.

Jill and I worked in the garden all summer. Our community grew many different vegetables. At the end of the summer, we picked enough to have a cookout. Mama brought a big pot of miso and vegetable stew. Everyone thanked Mama and Papa for their help. They brought a bit of Japan to Portland. I was so proud.

Look what we picked!

Make Connections

What did Kiku's family do to help their new community? What parts of their culture did they share? **ESSENTIAL QUESTION**

How can you and your family contribute to your community? **TEXT TO SELF**

Ask and Answer Questions

Ask yourself questions as you read. Then look for details to support your answers.

Find Text Evidence

Look at the section "Spring in the City" on page 55. Think of a question. Then read to answer it.

page 55

Spring in the City

My name is Kiku Sato. Last spring, my family and I moved from the country to the big city.

Our new home in Portland had no yard. There wasn't even a tiny plot of land. So Mama made an indoor garden. First she and Papa planted seeds in pots. Then they hung them from hooks. Next they crammed plants onto shelves. Green vines **tumbled** over desks. Soon our house had plants everywhere.

At first I was **scared** to start school. I was afraid no one would be my friend. But I soon met a **classmate**. Jill Hernandez and I were **practicing** reading aloud one day. She helped me say her last name, and I helped her **pronounce** mine. The next day we were best friends. Jill spent lots of time at my house.

I have a question. Why were there so many plants in Kiku's house? I read that they did not have a yard. Mama and Papa planted lots of seeds. I can answer my question. Kiku's family liked to grow things and didn't have the space to do it outdoors.

COLLABORATE

Your Turn

Reread "An Idea for a Garden." Think of one question. You might ask: Why did Jill's mother ask Kiku's mother for help? Read the section again to find the answer.

Sequence

Sequence is the order in which important events take place. Look for words, such as *first*, *next*, *then*, and *finally*. These signal words show the sequence of events.

 Find Text Evidence

In this autobiography, the events are told in sequence. I see the signal word first *in "Spring in the City" on page 55. I will read to find out what happens next. I will look for signal words to help me.*

Event
First Mama and Papa planted lots of seeds.

↓

Event
Then they hung pots from hooks.

↓

Event

Your Turn COLLABORATE

Reread "Spring in the City." What happens next? List the events in order in your graphic organizer.

Go Digital!
Use the interactive graphic organizer

61

Narrative Nonfiction

"Room to Grow" is an autobiography. An **autobiography**:

- Is a kind of narrative nonfiction
- Tells the true story of a person's life in order
- Is written by that person and uses *I* and *me*

 Find Text Evidence

I know "Room to Grow" is an autobiography. It is a true story written by Kiku about her life. She uses the words I *and* me. *Kiku's story also has headings and a map.*

page 55

Spring in the City

My name is Kiku Sato. Last spring, my family and I moved from the country to the big city.

Our new home in Portland had no yard. There wasn't even a tiny plot of land. So Mama made an indoor garden. First she and Papa planted seeds in pots. Then they hung them from hooks. Next they crammed plants onto shelves. Green vines **tumbled** over desks. Soon our house had plants everywhere.

At first I was **scared** to start school. I was afraid no one would be my friend. But I soon met a **classmate**. Jill Hernandez and I were **practicing** reading aloud one day. She helped me say her last name, and I helped her **pronounce** mine. The next day we were best friends. Jill spent lots of time at my house.

A Map Of Oregon

55

Text Features

Headings Headings tell what a section of text is mostly about.

Map A map is a flat drawing of a real place.

COLLABORATE

Your Turn

Find parts of "Room to Grow" that show you it is an autobiography. Tell your partner what you learned about Kiku and her culture.

Compound Words

A compound word has two small words in it. Put together the meanings of the two smaller words. Figure out the meaning of the compound word.

 Find Text Evidence

I see the compound word afternoon *on page 56. It has two smaller words,* after *and* noon. *I know what* after *means. I know* noon *means 12 o'clock. I think* afternoon *means a time after 12 o'clock.*

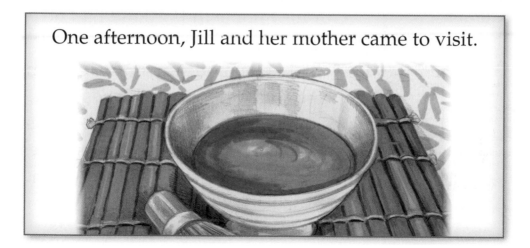

One afternoon, Jill and her mother came to visit.

Your Turn

COLLABORATE

Figure out the meanings of these compound words.
sunlight, *page 58, "A Garden Grows"*
cookout, *page 59, "A Garden Grows"*

Margaret Lindmark

Readers to...

Writers share their ideas in order and use words such as *first*, *next*, *then*, and *last*. Reread this passage from "Room to Grow." Think about how the writer organizes ideas.

Sequence

Name two words that show order. How do these words help you understand the way the story is organized?

Expert Model

Spring in the city

Our new home in Portland had no yard. There wasn't even a tiny plot of land. So Mama made an indoor garden. First she and Papa planted seeds in pots. Then they hung them from hooks. Next they crammed plants onto shelves. Green vines tumbled over desks. Soon our house had plants everywhere.

Writers

Ed wrote about something that happened to him. Read his revisions.

Editing Marks

≡ Make a capital letter.

/ Make a small letter.

⊙ Add a period.

∧ Add.

⌒ Take out.

Grammar Handbook

Subjects See page 475.

Student Model

My Life

When I was young, we live~~d~~ ᵉᵈ∧ with

First
my grandmother. ∧She lived in

Mexico. ~~t~~hen she moved to Texas.

I went to school there. My family

liked Texas⊙ ∧Now we have our own

house We have a big yard. My

lives with us
grandmother. ∧

By Ed H.

TEXAS

Your Turn   COLLABORATE

☑ Identify words that show order.
☑ Identify a subject.
☑ Tell how revisions improved the writing.

Go Digital!
Write online in Writer's Workspace

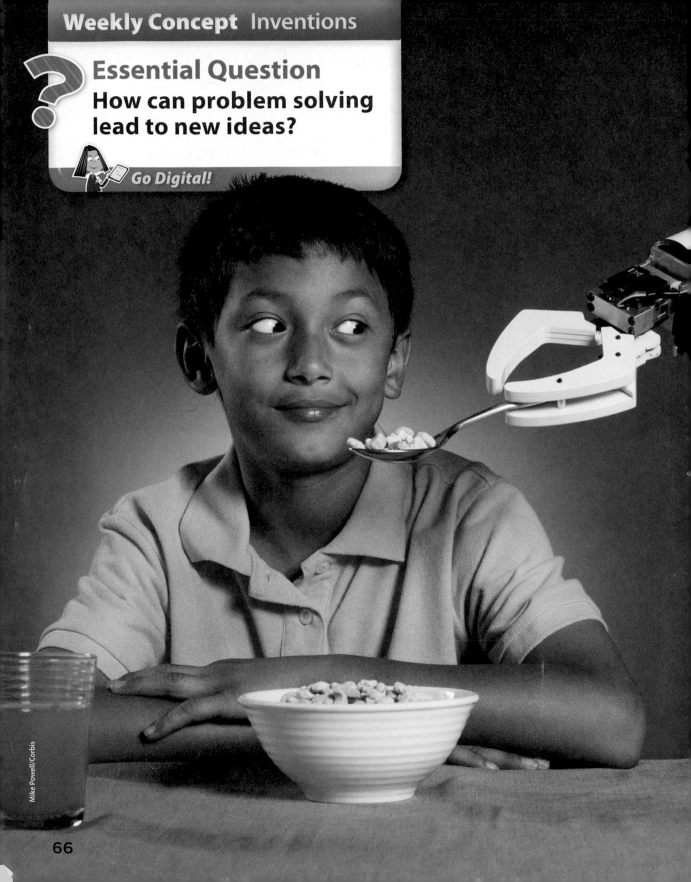

Essential Question
How can problem solving lead to new ideas?

Go Digital!

Mike Powell/Corbis

Inventing Something New

Almost every new invention starts with a problem and a new idea.

- ▶ Inventions make our lives better.
- ▶ Some inventions help us learn.
- ▶ Other inventions entertain us.

Talk About It

Talk with a partner about inventions. Write words that tell how inventors think up new ideas.

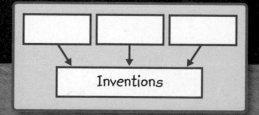

Inventions

Vocabulary

Use the picture and the sentence to talk with a partner about each word.

design

Jake and Andy are helping to **design** a picture for the school's new wall.

How would you design something new?

encouraged

My teacher **encouraged** us to eat healthy snacks.

Name something someone has encouraged you to do.

examine

The vet will **examine** my dog to make sure he is healthy.

What does the word examine mean?

investigation

Tom did a careful **investigation** of the spider web in his yard.

What can you do during an investigation?

quality

A good **quality** helmet helps keep me safe.

Why do people buy good quality things?

simple

Walking my dog is a **simple** and easy way to have fun.

What word in the sentence means the same as simple?

solution

Marco found a **solution** that helped him get better grades.

What does it mean to find a solution?

substitutes

Fruit and popcorn are healthful **substitutes** for candy.

Name two more healthful substitutes for sweets.

COLLABORATE

Your Turn

Pick three words. Then write three questions for your partner to answer.

Go Digital! *Use the online visual glossary*

Mary Anderson's GREAT INVENTION

? Essential Question

How can problem solving lead to new ideas?

Read about how someone solved a problem and invented something new.

You might think that a ride in a bus or car is the same today as it was long ago. That isn't true. The first cars were not as fast. They were noisy. Cars didn't even have windshield wipers!

When it rained, drivers rubbed their windshields with an onion. The oil from the onion would repel, or keep off, rain and sleet. It wasn't the best **solution**, but there were no better **substitutes**. Nothing else worked. Then a woman named Mary Anderson solved the problem.

It Started with Snow

Mary Anderson grew up in Alabama. In the winter of 1902, she went to New York City. It was a cold and windy day. The sky was a gray curtain. Snow was a white blanket on the ground. Mary was cold and wet. Because she wanted to warm up and get dry, she rode a streetcar.

Back then, some streetcar windshields had two parts. They opened with a push. From her seat, Mary watched snow and ice build up on the windshield. The streetcar driver could not see. So, he pushed open the windshield. This helped him to see better. As a result, snow and ice blew in his face. Soon his nose and ears were ice cubes.

Other cars kept stopping, too. Sometimes the drivers hopped out. They wiped off their windshields. Then, they got back in and drove. As a result, traffic moved slowly.

Malene Laugesen

The Next Step

Mary thought about this problem. How could drivers clean their windshields without stopping? Could they do it without opening their windshield?

Back home in Alabama, Mary sketched her idea. Then she added notes. She wanted to **examine** her solution to make sure that it worked. Next, Mary did her own **investigation**. She looked for facts about what drivers needed. She invented a windshield wiper that a driver could use from inside the car. Then she worked out a **design**, or plan. On paper, Mary's invention looked **simple**. She hoped drivers would find it easy to use.

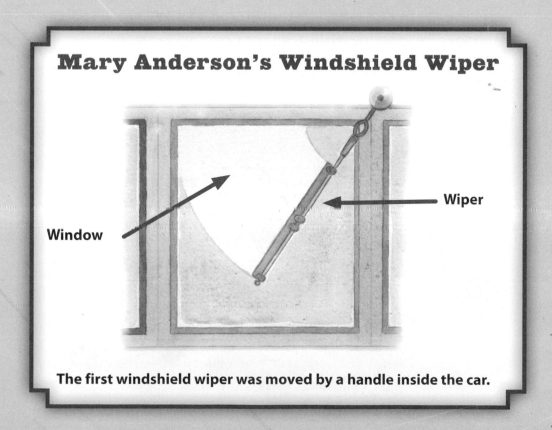

Mary Anderson's Windshield Wiper

Window

Wiper

The first windshield wiper was moved by a handle inside the car.

Mary had a model built. It was made of **quality** wood, rubber, and metal. Soon the model was ready to test. It was fitted on a windshield. The driver moved a handle inside the car. The handle caused a blade to move back and forth across the glass. It worked! Mary's idea was a gem! She felt **encouraged** and was sure it would sell.

Solving the Problem

Mary's windshield wipers solved a problem. But it took many years before people used them. That's because most people did not own cars.

By 1913, more people bought and drove cars. Those cars had windshields. Finally windshield wipers began to sell. Driving became safer and easier because of Mary Anderson's idea.

Safer to Drive

Cars from long ago were different from cars we ride in today. Here are some more inventions that helped make driving safer.

- The first seat belts were used in 1885.
- Cars stopped at the first stop sign in 1915.
- Cars first used turn signals in 1938.

Make Connections

Talk about how Mary Anderson's solution to a problem led to a new idea. **ESSENTIAL QUESTION**

What inventions can you think of that have made your life better? **TEXT TO SELF**

Malene Laugesen

Ask and Answer Questions

Ask yourself questions as you read "Mary Anderson's Great Invention." Then look for details to support your answers.

 Find Text Evidence

Look at the section "It Started with Snow" on page 72. Think of a question. Then read to answer it.

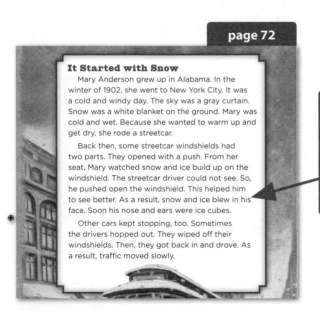

page 72

It Started with Snow

Mary Anderson grew up in Alabama. In the winter of 1902, she went to New York City. It was a cold and windy day. The sky was a gray curtain. Snow was a white blanket on the ground. Mary was cold and wet. Because she wanted to warm up and get dry, she rode a streetcar.

Back then, some streetcar windshields had two parts. They opened with a push. From her seat, Mary watched snow and ice build up on the windshield. The streetcar driver could not see. So, he pushed open the windshield. This helped him to see better. As a result, snow and ice blew in his face. Soon his nose and ears were ice cubes.

Other cars kept stopping, too. Sometimes the drivers hopped out. They wiped off their windshields. Then, they got back in and drove. As a result, traffic moved slowly.

I have a question. Why did drivers push open their windshields? <u>I read that windshields had two parts. The driver pushed it open. This helped him see where he was going.</u> Now I can answer my question. Opening the windshield helped the driver see better.

Your Turn

Think of one question about Mary Anderson's invention. You might ask: How did it work? Reread page 74 to answer it.

Malene Laugesen

Cause and Effect

A cause is why something happens. An effect is what happens. They happen in time order. Signal words, such as *because* and *as a result*, show cause and effect.

 Find Text Evidence

On page 72 I read that Mary rode a streetcar. This is the effect. Now I can find the cause. Mary was cold and wet. She wanted to warm up and get dry. The signal word because *helped me find the cause and effect.*

Cause		Effect
First: Mary was cold and wet and wanted to warm up.	➡	Mary rode a streetcar.
Next Snow and ice built up on the streetcar's windshield.	➡	

Your Turn

COLLABORATE

Reread "Mary Anderson's Great Invention." Use signal words to find more causes and effects. Fill in the graphic organizer. Make sure events are in time order.

Go Digital!
Use the interactive graphic organizer

Biography

"Mary Anderson's Great Invention" is a biography.
A **biography**:
- Tells the true story of a real person's life and is written by another person
- Is told in sequence, or time order
- May have diagrams or sidebars

 Find Text Evidence

"Mary Anderson's Great Invention" is a biography. It is a true story about a person's life. It is written by another person. The events are told in order.

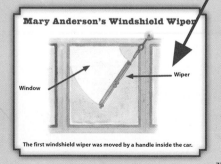

page 73

The Next Step

Mary thought about this problem. How could drivers clean their windshields without stopping? Could they do it without opening their windshield?

Back home in Alabama, Mary sketched her idea. Then she added notes. She wanted to **examine** her solution to make sure that it worked. Next, Mary did her own **investigation**. She looked for facts about what drivers needed. She invented a windshield wiper that a driver could use from inside the car. Then she worked out a **design**, or plan. On paper, Mary's invention looked **simple**. She hoped drivers would find it easy to use.

Mary Anderson's Windshield Wiper

Window — Wiper

The first windshield wiper was moved by a handle inside the car.

73

Text Features

Diagrams Diagrams are simple drawings with labels.

Sidebars Sidebars give more information about a topic.

 COLLABORATE

Your Turn

Reread "Mary Anderson's Great Invention." Find text features. Tell what you learned from each of the text features.

Metaphors

"The sun is a yellow ball" is a metaphor. A metaphor compares two things that are very different. Look for metaphors as you read.

Find Text Evidence

On page 72, I see the sentence "The sky was a gray curtain." This is a metaphor. It compares the sky to a gray curtain. It means that the sky was dark and cloudy. This metaphor helps me picture a dark and gloomy sky.

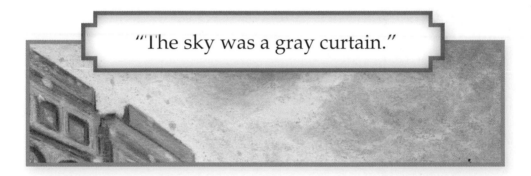

"The sky was a gray curtain."

Your Turn

COLLABORATE

Talk about these metaphors from the story. What two things do they compare?

Snow was a white blanket on the ground.
page 72

Soon his nose and ears were ice cubes.
page 72

Malene Laugesen

Readers to...

Writers put their ideas in order. They choose words that show when things happen. They use words such as *first*, *next*, *then*, and *last*. Reread this passage from "Mary Anderson's Great Invention."

Time-Order Words

Find words that show when things happen. How do these words help put ideas in order?

Expert Model

Back home in Alabama, Mary sketched her idea. Then she added notes. She wanted to examine her solution to make sure that it worked. Next, Mary did her own investigation. She looked for facts about what drivers needed. She invented a windshield wiper that a driver could use from inside the car.

Malene Laugesen

Writers

Jim wrote about an inventor. Read his revisions.

Editing Marks

≡ Make a capital letter.
/ Make a small letter.
⊙ Add a period.
∧ Add
⌐ Take out.

Grammar Handbook

Predicates See page 475.

Student Model

Garrett Morgan
was an inventor.

Garret Morgan. He invented ∧

things that made people safe? ⊙ ∧

First he invented the traffic signal.

Then
∧He invented a mask. People wear

his masks to breathe fresh air.

They wear them to fight fires. we ≡

still use his inventions today.

by Jim F.

Your Turn

COLLABORATE

- ☑ Identify time order words.
- ☑ Find a predicate.
- ☑ Tell how revisions improved the writing.

Go Digital!
Write online in Writer's Workspace

Essential Question

How do landmarks help us understand our country's story?

Go Digital!

UNDERSTANDING HISTORY

Martin Luther King, Jr. played an important role in American history. This monument in Washington, D.C. honors his life.

▶ People visit monuments to learn about important people and events in history.

▶ Landmarks and monuments help us remember and understand history.

Talk About It

Write words you have learned about landmarks. Talk with a partner about how these monuments help people learn about history.

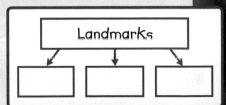

Landmarks

Vocabulary

Use the picture and the sentence to talk with a partner about each word.

carved

A strong river **carved** this canyon out of rock.

What other things can be carved?

clues

These paw prints are **clues** that a dog walked here today.

What clues tell you that it might rain?

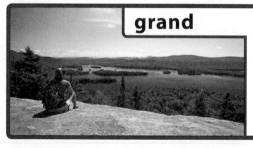

grand

Diane sat and gazed at the **grand** view of the river.

What do you think would make a grand view?

landmark

The Statue of Liberty is an American **landmark**.

What other landmarks can you name?

massive

The boaters looked up at the **massive** stone cliff.

What is another word for massive?

monument

This **monument** honors the leaders of our country.

Describe a monument you have seen.

national

The Fourth of July is a **national** holiday.

Name another national holiday.

traces

In the morning we found **traces**, or small amounts, of snow on the plants.

What is another word for traces?

Your Turn

COLLABORATE

Pick three words. Write three questions for your partner to answer.

Go Digital!   *Use the online visual glossary*

Essential Question

How do landmarks help us understand our country's story?

Read about what one national landmark teaches us.

Kristy-Anne Glubish/Design Pics

A Natural Beauty

It is a famous **landmark** in the United States, and it's huge! It is one mile deep and ten miles wide. It was **carved** out of rock by the Colorado River. It stretches across parts of four states. What is it? It's the **Grand** Canyon!

Exploring the Canyon

Many tourists visit the Grand Canyon. In fact, almost five million people take a trip to this **national** treasure each year. People come from around the world to hike the dusty trails. They take boat rides down the Colorado River. They gaze across miles of **massive** red and brown cliffs.

Nature lovers visit the Grand Canyon, too. They come to look for animals. They peek at the hundreds of different kinds of plants. They may spot eagles and see mountain lions. They may spy snakes and spiders, and some may even see bats. Some visitors also come to learn about the canyon's history.

History of the Canyon

Explorers from Europe first saw the Grand Canyon in 1540. Then in 1857, American explorers discovered it. They found groups of Native Americans living there. One of these groups was the Ancient Pueblo people.

The Ancient Pueblo people lived in the canyon for almost one thousand years. They were farmers and hunters. Scientists have found **traces**, or parts, of their old homes still standing.

The Ancient Pueblo people lived in cliff houses like these.

Scientists have also found very old rocks in the Grand Canyon. These rocks are some of the oldest in the world. They are clues to how the canyon was formed. Some scientists look for **clues** about the people who lived there. They have found tools and pieces of pottery.

A Great Big Park

This map shows where the Grand Canyon is located.

UTAH

NEVADA

GRAND CANYON NATIONAL PARK

Colorado River

Lake Mead

Las Vegas

15

Colorado River

North Rim

Grand Canyon Village

○ City
— Highway
▨ Grand Canyon National Park

ARIZONA

N
W   E
S

Kingman

40

Flagstaff

It's a Landmark

President Theodore Roosevelt visited the Grand Canyon in 1903. He saw how beautiful it was. He said it was a special place. As a result, he made it a national **monument**. Then in 1919, the Grand Canyon was declared a national park. That means the land is protected. No one can build homes on it. The Grand Canyon is a place all Americans can enjoy.

Protect the Canyon

It is important for people to take care of national landmarks. We can do our part by following the rules when we visit. Animals and wildlife are safe there and should not be touched. Rivers must be kept clean.

There is still a lot to learn about this beautiful landmark. It is important that we protect it.

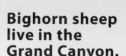

Bighorn sheep live in the Grand Canyon.

Make Connections

How does the Grand Canyon teach us about America's story? **ESSENTIAL QUESTION**

What do you find most interesting about the Grand Canyon's history? Why? **TEXT TO SELF**

Matt Dil/Flickr/Getty Images

Ask and Answer Questions

Stop and ask yourself questions as you read. Then look for details to support your answers.

 Find Text Evidence

Reread the section "Exploring the Canyon" on page 87. Think of a question. Then read to answer it.

> page 87
>
> ### Exploring the Canyon
>
> Many tourists visit the Grand Canyon. In fact, almost five million people take a trip to this **national** treasure each year. People come from around the world to hike the dusty trails. They take boat rides down the Colorado River. They gaze across miles of **massive** red and brown cliffs.
>
> Nature lovers visit the Grand Canyon, too. They come to look for animals. They peek at the hundreds of different kinds of plants. They may spot eagles and see mountain lions. They may spy snakes and spiders, and some may even see bats. Some visitors also come to learn about the canyon's history.
>
> 87

I have a question. Why do people visit the Grand Canyon? I read that people like to hike trails and take boat rides. They like to look at the animals and cliffs. They go to learn about its history. Now I can answer my question. People visit the canyon for many reasons.

Your Turn

COLLABORATE

Reread "History of the Canyon." Think of one question. You might ask: How do we know about people who once lived in the canyon? Read the section again to find the answer.

Kristy-Anne Glubish/Design Pics

Main Idea and Key Details

The main idea is the most important point the author makes about a topic. Key details tell about the main idea.

 Find Text Evidence

What details tell about why people visit the Grand Canyon? I can reread page 87 and find important details. Then I can figure out what these key details have in common to tell the main idea.

Main Idea
Detail
People hike the trails and take boat rides down the river.
Detail
They like to gaze across the massive red and brown cliffs.
Detail

Your Turn

Reread. Find more key details about why people visit the Grand Canyon. List them in your graphic organizer. Then use the details to figure out the main idea.

Go Digital!
Use the interactive graphic organizer

Expository Text

"A Natural Beauty" is an expository text.

Expository text:
- Gives facts and information about a social studies topic
- Includes text features such as photographs, captions, sidebars, and maps

 Find Text Evidence

I can tell that "A Natural Beauty" is an expository text. It gives facts and information about the Grand Canyon. It also has photographs, captions, a sidebar, and a map.

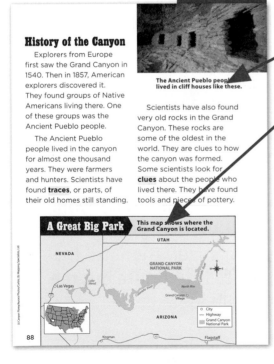

page 88

History of the Canyon

Explorers from Europe first saw the Grand Canyon in 1540. Then in 1857, American explorers discovered it. They found groups of Native Americans living there. One of these groups was the Ancient Pueblo people.

The Ancient Pueblo people lived in the canyon for almost one thousand years. They were farmers and hunters. Scientists have found **traces**, or parts, of their old homes still standing.

The Ancient Pueblo people lived in cliff houses like these.

Scientists have also found very old rocks in the Grand Canyon. These rocks are some of the oldest in the world. They are clues to how the canyon was formed. Some scientists look for **clues** about the people who lived there. They have found tools and pieces of pottery.

A Great Big Park

This map shows where the Grand Canyon is located.

UTAH
NEVADA
GRAND CANYON NATIONAL PARK
Las Vegas
Lake Mead
North Rim
Grand Canyon Village
ARIZONA
○ City
— Highway
Grand Canyon National Park
Kingman
Flagstaff

88

Text Features

Captions Captions give extra information that is not in a text.

Map A map is a flat drawing of a real place.

Sidebar A sidebar gives more information about a topic.

COLLABORATE

Your Turn

Find more text features in "A Natural Beauty." What else did you learn?

Multiple-Meaning Words

Multiple-meaning words have more than one meaning. Find other words in the sentence to help you figure out the meaning of a multiple-meaning word.

 Find Text Evidence

On page 87 I see bats. *This word can mean "wooden sticks used to hit a ball" or "small animals that fly." The context clues, "spy snakes and spiders" can help me figure out that* bats *are animals in this sentence. Now I know that* bats *here are "small animals that fly at night."*

They may spy snakes and spiders, and some may even see bats.

Your Turn

 COLLABORATE

Use context clues to figure out the meanings of the following words.

spot, *page 87*
safe, *page 89*

Kristy-Anne Glubish/Design Pics

Readers to...

Writers often use different kinds of sentences in their writing. Statements, questions, and exclamations make their writing more interesting to read. Reread this passage from "A Natural Beauty."

Sentence Types

With a partner, identify three different **sentence types** the author uses. How do they make the story more interesting to read?

Expert Model

It is a famous landmark in the United States, and it's huge! It is one mile deep and ten miles wide. It was carved out of rock by the Colorado River. It stretches across parts of four states. What is it? It's the Grand Canyon!

Writers

Ron wrote about why it is important to have parks. Read his revision.

Student Model

OUR PARKS ARE IMPORTANT

Don't you think national parks
are an important part of our
?
country. People visit parks to
, and
see animals. They go to have fun.
in a park.
They can walk, bike, and play. It is
important to keep our parks open.
Everyone should be able to
s
go to national park.

By Ron H.

Editing Marks

- Make a capital letter.
- Make a small letter.
- Add a period.
- Add.
- Take out.

Grammar Handbook

Simple and Compound Sentences See page 476.

Your Turn

- ✔ Identify different types of sentences.
- ✔ Identify simple and compound sentences.
- ✔ Tell how revisions improved the writing.

Go Digital!
Write online in Writer's Workspace

Unit 2

Figure It Out

The Big Idea
What does it take to solve a problem?

The Problem Solver

If you have a problem,
    Super-sized or small,
I've got some suggestions,
    Just give me a call.

Do a lot of reading,
    Study every fact,
Write down some solutions,
    Think before you act.

If the problem is too hard,
    Ask a friend or two,
Teamwork's always better,
    When there's work to do.

Problems always pop up,
    Do not scream or cry,
You'll find a solution,
    If you only try.

— Constance Andrea Keremes

? Essential Question

Why is working together a good way to solve a problem?

Go Digital!

Leland Bobbé/Fuse/Getty Images

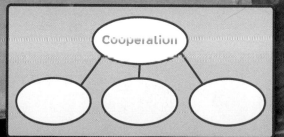

WORKING TOGETHER

There was a problem, and these friends figured out what to do. Now they are working together to solve it.

▶ Working together is a good way to solve problems.

▶ Cooperation makes the job easier.

Talk About It

Write words you have learned about working together. Talk with a partner about how cooperation helps get things done.

Vocabulary

Use the picture and the sentence to talk with a partner about each word.

attempt

Maria made an **attempt** to run a mile.

When have you made an attempt to do something new?

awkward

The penguin looked **awkward** and clumsy.

What word in the sentence means the same as awkward?

cooperation

I cleaned the floor faster with Becky's **cooperation**.

Tell about a time when cooperation made a job easier.

created

Jim and his grandfather **created** a home for birds out of wood.

What is another word for created?

furiously

The rattlesnake shook its tail **furiously** as a warning.

What might make a rattlesnake shake its tail furiously?

interfere

The rain is going to **interfere** with our ballgame.

What kind of weather might interfere with your plans?

involved

Our class is **involved** in the school play.

What activity are you involved in?

timid

The shy, **timid** kitten hid under a blanket.

What is the opposite of timid?

COLLABORATE

Your Turn

Pick three words. Then write three questions for your partner to answer.

Go Digital! *Use the online visual glossary*

Anansi
Learns a Lesson

? Essential Question

Why is working together a good way to solve a problem?

Read how Turtle works with a friend to solve his problem.

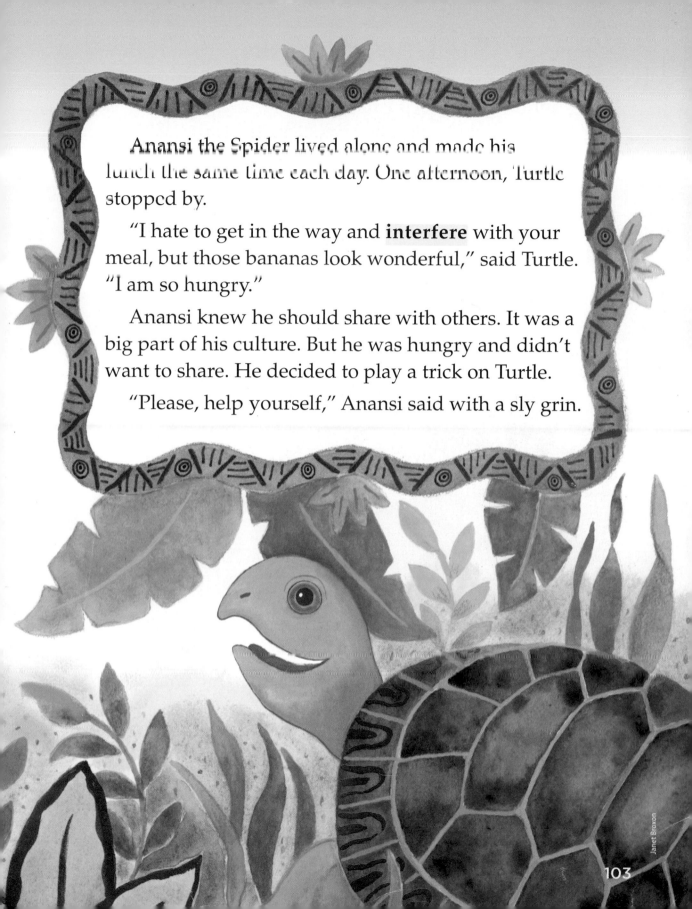

Anansi the Spider lived alone and made his lunch the same time each day. One afternoon, Turtle stopped by.

"I hate to get in the way and **interfere** with your meal, but those bananas look wonderful," said Turtle. "I am so hungry."

Anansi knew he should share with others. It was a big part of his culture. But he was hungry and didn't want to share. He decided to play a trick on Turtle.

"Please, help yourself," Anansi said with a sly grin.

Janet Broxon

Turtle reached for the food. "Shouldn't you wash your hands?" asked Anansi.

"Oh, yes!" Turtle said. When Turtle returned, Anansi had eaten half of the bananas.

"I didn't want the bananas to spoil," said Anansi.

Turtle got closer and made another **attempt** to eat. Anansi stopped him again.

"Turtle, please go wash your hands," he said.

Turtle knew his hands were clean, but Anansi still thought they were filthy. However, Turtle was too shy and **timid** to say no. When he returned Anansi had eaten the rest of the fruit.

"Ha, ha, I tricked you, Turtle," said Anansi. "You didn't get any bananas!"

Janet Broxon

Turtle was angry at Anansi. He decided to teach that nasty spider a lesson. "Please come to my house at the bottom of the lake for dinner tomorrow," said Turtle.

Anansi quickly said yes. He loved free food.

Turtle couldn't trick Anansi alone, so he decided to ask Fish to get **involved** and help make a plan.

Turtle found Fish at the lake. "Fish, I need your help," he said. "With your **cooperation**, we can trick Anansi." Anansi had tricked Fish many times so Fish was happy to help. Together the two friends **created** a clever plan.

The next day, Anansi went to the lake. Fish met him at the water's edge. "Come Anansi," said Fish. "We will swim to Turtle's house together." Anansi jumped into the water. He was a clumsy and **awkward** swimmer. He was also very light.

"How will I ever get down to Turtle's house?" he cried.

Fish knew what to say. "Grab some heavy stones. Then you will sink, not float."

Anansi picked up two big stones, jumped into the lake, and sank down, down, down. Fish swam at his side. At Turtle's house, Anansi saw a wonderful feast of berries.

Janet Brown

"Welcome, Anansi," said Turtle. "Drop those stones and help yourself."

As soon as Anansi dropped the stones, he rocketed to the surface of the lake. Anansi sputtered **furiously**. "Fish and Turtle tricked me," he cried angrily.

Back at the bottom of the lake, Turtle and Fish laughed and laughed.

"We worked together and taught Anansi a lesson," said Turtle.

"What a good way to solve a problem," said Fish. "Let's eat!"

Make Connections

Tell how Turtle and Fish worked together to trick Anansi. **ESSENTIAL QUESTION**

Think of a time when you and a friend solved a problem. Why was it easier to work together? **TEXT TO SELF**

Make Predictions

Use clues in the story to guess, or predict, what happens next. Was your prediction right? Read on to check it. Change it if it is not right.

 Find Text Evidence

You may have made a prediction about Anansi at the beginning of "Anansi Learns a Lesson." What clues on page 103 helped you guess what might happen?

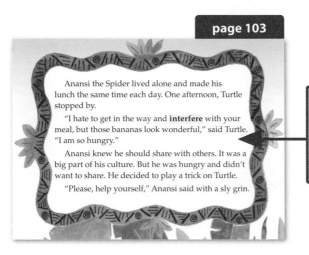

page 103

Anansi the Spider lived alone and made his lunch the same time each day. One afternoon, Turtle stopped by.

"I hate to get in the way and **interfere** with your meal, but those bananas look wonderful," said Turtle. "I am so hungry."

Anansi knew he should share with others. It was a big part of his culture. But he was hungry and didn't want to share. He decided to play a trick on Turtle.

"Please, help yourself," Anansi said with a sly grin.

I predicted that Anansi would eat the bananas. I read that <u>Anansi was hungry and didn't want to share. He decided to play a trick on Turtle.</u> I will read on to check my prediction.

Your Turn

COLLABORATE

What prediction did you make after Turtle asked Fish for help? Reread to check your prediction. Remember to make, confirm, and revise predictions as you read.

Janet Broxon

Theme

The theme of a story is the author's message. Think about what the characters do and say. Use these key details to help you figure out the theme.

 Find Text Evidence

In "Anansi Learns a Lesson," Turtle and Fish work together to solve a problem. This is the story's theme. I can reread to find key details that help me figure out the theme.

> **Detail**
> Anansi tricks Turtle so Turtle decides to trick Anansi.

↓

> **Detail**
> Turtle asks Fish to help, and they make a plan.

↓

> **Detail**

↓

> **Theme**
> Working together is a good way to solve problems.

Your Turn COLLABORATE

Reread "Anansi Learns a Lesson." List an important detail about Turtle's actions in your graphic organizer. Be sure the detail tells about the theme.

Go Digital!
Use the interactive graphic organizer

Folktale

"Anansi Learns a Lesson" is a folktale. A **folktale**:
- Is a short story passed from person to person in a culture
- Usually has a message or lesson

 Find Text Evidence

I can tell that "Anansi Learns a Lesson" is a folktale. I learned something about Anansi's culture. I also learned a lesson.

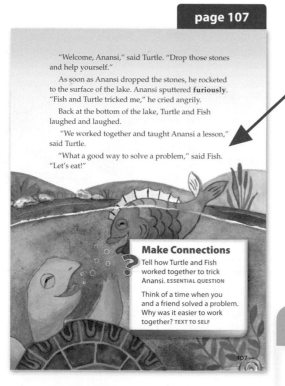

page 107

"Welcome, Anansi," said Turtle. "Drop those stones and help yourself."

As soon as Anansi dropped the stones, he rocketed to the surface of the lake. Anansi sputtered **furiously**. "Fish and Turtle tricked me," he cried angrily.

Back at the bottom of the lake, Turtle and Fish laughed and laughed.

"We worked together and taught Anansi a lesson," said Turtle.

"What a good way to solve a problem," said Fish. "Let's eat!"

Make Connections

Tell how Turtle and Fish worked together to trick Anansi. ESSENTIAL QUESTION

Think of a time when you and a friend solved a problem. Why was it easier to work together? TEXT TO SELF

107

The message or lesson in a folktale is sometimes found at the end of the story. It is a message the author thinks is important.

Your Turn

Think about the lesson that Anansi learns in this story. Tell your partner the message.

Antonyms

Antonyms are words that have opposite meanings. Look for antonyms to help you figure out the meaning of unknown words.

 Find Text Evidence

In "Anansi Learns a Lesson," I see the word filthy *in a sentence on page 104. I'm not sure what* filthy *means so I'll look for context clues. I see the word* clean. *I know what* clean *means. When I reread, I see that* clean *and* filthy *are antonyms. I think that* filthy *means "not clean."*

Turtle knew his hands were clean,
but Anansi still thought they were filthy.

Your Turn

 COLLABORATE

Find an antonym to help you figure out the meaning of this word.

sink, *page 106*

Talk about how the antonym helped you figure out what *sink* means.

Janet Broxon

Readers to . . .

Writers choose words to link, or connect, their ideas. They use words, such as *and, because, however,* and *but.* Reread this passage from "Anansi Learns a Lesson."

Linking Ideas

Find two words that connect ideas. Why are these words good examples of **word choice**?

Expert Model

Turtle knew his hands were clean, but Anansi still thought they were filthy. However, Turtle was too shy and timid to say no. When he returned Anansi had eaten the rest of the fruit.

"Ha, ha, I tricked you, Turtle," said Anansi. "You didn't get any bananas!"

Writers

Marta wrote about solving a problem. Read her revisions.

Editing Marks

≡ Make a capital letter.

／ Make a small letter.

⊙ Add a period

∧ Add

ℓ Take out.

Grammar Handbook

Common and Proper Nouns
See page 478.

Student Model

The Lost Book

Pam and I found a ~~B~~ook on the playground. Who did it belong to？ We had to find out. Pam wanted to make her own sign. ^But^ ~~W~~e made one together. We put up the sign at state street school. Our pal, mark, came and got his book.

by Marta H.

Book found. Call Pam.

Your Turn

- ✔ Identify words that link ideas.
- ✔ Find common and proper nouns.
- ✔ Tell how revisions improved the writing.

Go Digital!
Write online in Writer's Workspace

113

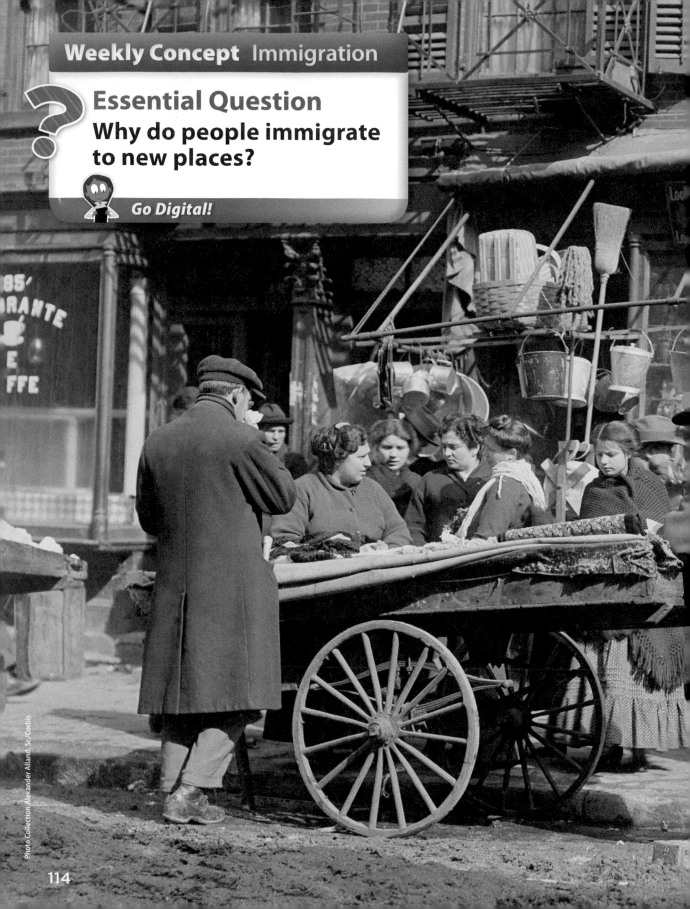

Essential Question

Why do people immigrate to new places?

Go Digital!

Finding a Home

Welcome to Hester Street. Many immigrants moved here and worked on this New York City street. They came for many reasons.

► Immigrants dreamed about new jobs.

► They felt there were lots of opportunities.

► They believed their lives would be better.

Talk About It

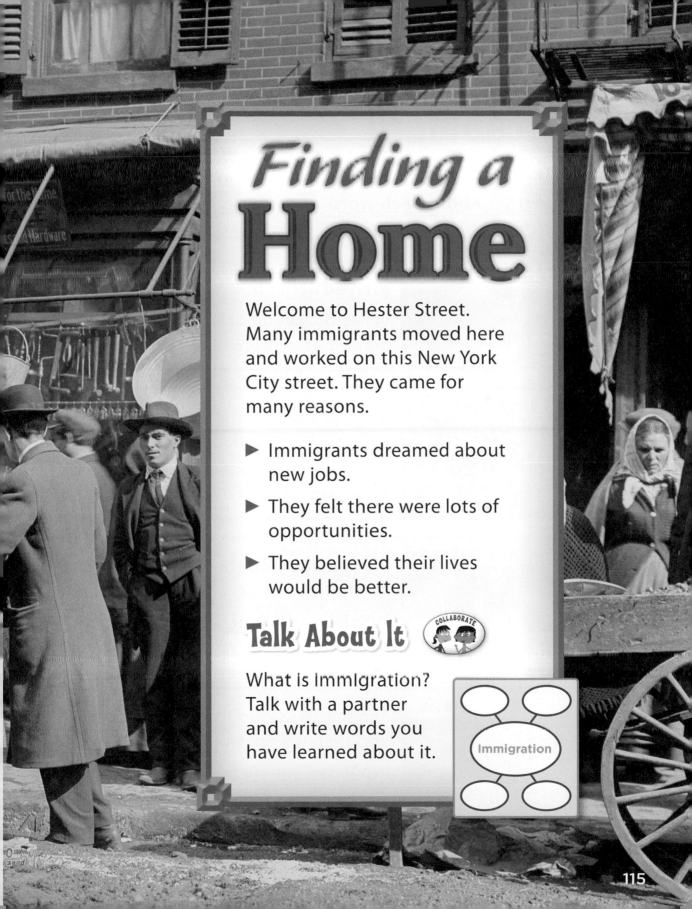 COLLABORATE

What is Immigration? Talk with a partner and write words you have learned about it.

Immigration

Vocabulary

Use the picture and the sentence to talk with a partner about each word.

arrived

Aunt Sophie was so happy when we **arrived** at her house.

What do you do after you have arrived at school?

immigrated

Many people **immigrated** to the United States from other countries.

Why have people immigrated to America?

inspected

Stacy **inspected** the pine cone carefully.

What is something you have inspected carefully?

moment

The match stayed lit for a **moment**, and then it went out.

What can you do in a moment?

opportunity

Our class had an **opportunity**, or chance, to visit the museum.

What is another word for opportunity?

photographs

I like to look at old family **photographs**.

Why do people take photographs?

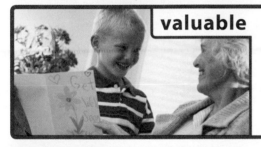

valuable

The card I made is very **valuable** to my grandmother.

Name something that is valuable to you.

whispered

Trudy **whispered** the plan to Josh so no one else could hear.

What is the opposite of whispered?

Your Turn

COLLABORATE

Pick three words. Then write three questions for your partner to answer.

Go Digital! *Use the online visual glossary*

SAILING TO AMERICA

Essential Question

Why do people immigrate to new places?

Read about why one family came to America.

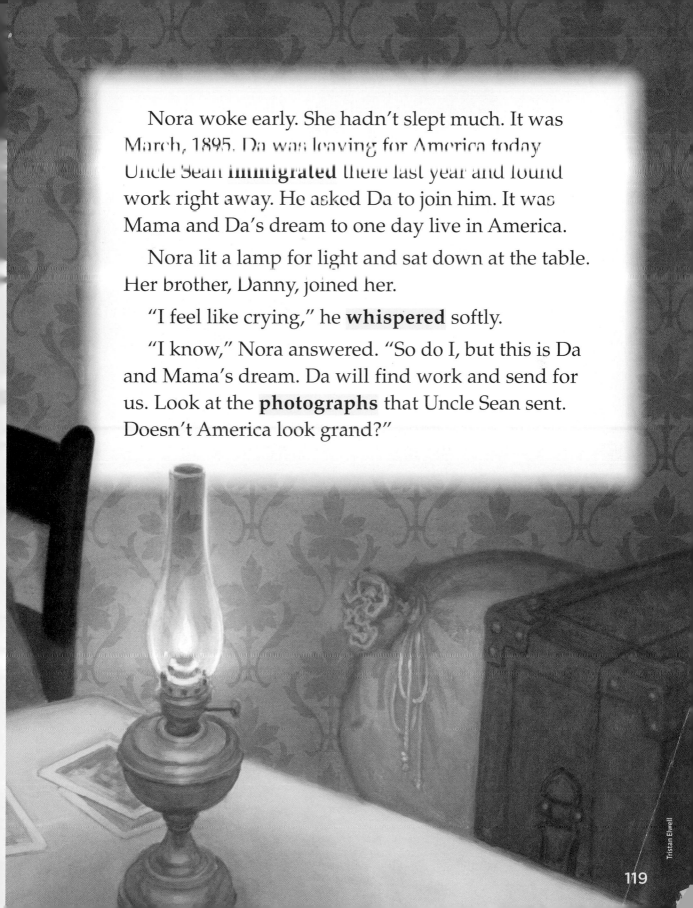

Nora woke early. She hadn't slept much. It was March, 1895. Da was leaving for America today. Uncle Sean **immigrated** there last year and found work right away. He asked Da to join him. It was Mama and Da's dream to one day live in America.

Nora lit a lamp for light and sat down at the table. Her brother, Danny, joined her.

"I feel like crying," he **whispered** softly.

"I know," Nora answered. "So do I, but this is Da and Mama's dream. Da will find work and send for us. Look at the **photographs** that Uncle Sean sent. Doesn't America look grand?"

Tristan Elwell

Tristan Elwell

"I don't want to ever leave Ireland," Danny said. "We won't have any friends in America. We'll be far away from Grandda, Paddy, and Colleen."

"Maybe you'll be glad it isn't Ireland," Nora said. "There will be enough food to eat. Mama and Da can relax and not worry so much. We'll all have a better life. America will be the land of our dreams."

Then Da carried a bag into the room. "Cheer up, my little loves! Why, in no time at all, you'll be joining me."

A year later, Da had saved enough money to send for his family. Mama, Danny, and Nora packed what little they had. They got on a crowded steamship and began their voyage.

The trip across the Atlantic Ocean was rough. The air inside the steamship smelled like a dirty sock. The ship tossed up and down for days. The waves were as big as mountains. Many passengers became seasick, but Nora and Danny felt fine.

Every day Nora daydreamed and reread Da's letters. She thought of the buildings and streetcars he wrote about. In her dreams, she could picture Da on a crowded street. He had a big smile on his face.

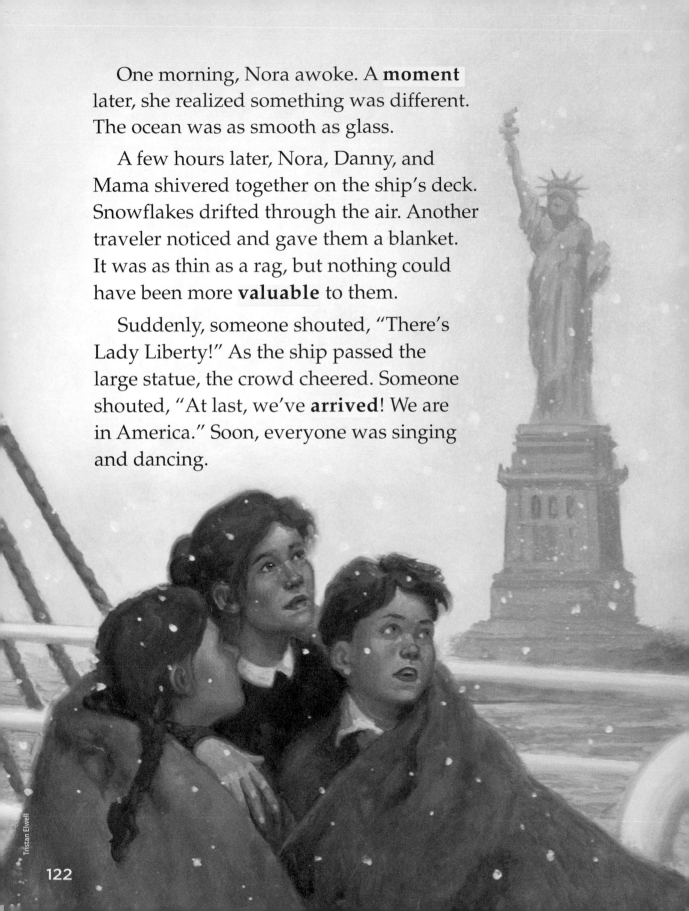

One morning, Nora awoke. A **moment** later, she realized something was different. The ocean was as smooth as glass.

A few hours later, Nora, Danny, and Mama shivered together on the ship's deck. Snowflakes drifted through the air. Another traveler noticed and gave them a blanket. It was as thin as a rag, but nothing could have been more **valuable** to them.

Suddenly, someone shouted, "There's Lady Liberty!" As the ship passed the large statue, the crowd cheered. Someone shouted, "At last, we've **arrived**! We are in America." Soon, everyone was singing and dancing.

Tristan Elwell

A ferry took the travelers to Ellis Island. In the main hall, doctors **inspected** the family. They looked for signs of illness. Mama had to answer many questions. Nora knew that people didn't get an **opportunity**, or another chance, to take these tests twice. Nora looked at Danny, then at Mama. They had to pass.

After a few hours, the family learned they could stay in America. As they filed off the ferry, Nora saw Uncle Sean's dark hair. Then she saw Da. His hands waved wildly. He had a big smile on his face. Dreams do come true, Nora thought as she waved back.

Make Connections

? Why did Nora and her family immigrate to America? How would the move make their lives better?

ESSENTIAL QUESTION

Has anyone in your family ever moved to a new place? How did they feel?

TEXT TO SELF

Make Predictions

Use story clues to predict what happens next. Was your prediction right? Reread to confirm, or check, it. Change it if it is not right.

 ### Find Text Evidence

What would Da do after he left? You may have predicted that he would send for his family. Reread pages 120–121 for clues to support your prediction.

page 120

"I don't want to ever leave Ireland," Danny said. "We won't have any friends in America. We'll be far away from Grandda, Paddy, and Colleen."

"Maybe you'll be glad it isn't Ireland," Nora said. "There will be enough food to eat. Mama and Da can relax and not worry so much. We'll all have a better life. America will be the land of our dreams."

Then Da carried a bag into the room. "Cheer up, my little loves! Why, in no time at all, you'll be joining me."

I predicted that Da would bring his family to America. Here is the clue. Da says they will be joining him. I read page 121 to check it. A year later, Da saved enough money to send for his family.

COLLABORATE

Your Turn

Predict what will happen when the family arrives in America. Find clues to support your prediction. Remember to make, confirm, and revise predictions as you read.

Tristan Elwell

Theme

The theme of a story is the author's message. Think about what the characters do and say. Use these key details to help you figure out the theme.

 Find Text Evidence

In "Sailing to America," Mama and Da dreamed of living in America. I think this is an important detail about the theme. I will reread to find more key details. Then I can figure out the story's theme.

Detail
It was Mama and Da's dream to live in America. Pa goes first.

↓

Detail
Nora tells Danny that America will be the land of their dreams.

↓

Detail

↓

Theme

Details tell what the characters do and say. They help you figure out the theme.

Your Turn COLLABORATE

Reread "Sailing to America." Find more important details then use them to figure out the theme. Write them in your graphic organizer.

Go Digital!
Use the interactive graphic organizer

125

Historical Fiction

"Sailing to America" is historical fiction. **Historical fiction:**

- Is a made-up story that takes place in the past
- Has illustrations that show historical details

🔍 Find Text Evidence

I can tell that "Sailing to America" is historical fiction. The characters and events are made up. The story is based on real events that happened a long time ago.

page 119

Nora woke early. She hadn't slept much. It was March, 1895. Da was leaving for America today. Uncle Ed had **immigrated** there last year and found work right away. He had asked Da to join him. It was Mama and Da's dream to one day live in America.

Nora lit a lamp for light and sat down at the table. Her brother, Danny, joined her.

"I feel like crying," he **whispered** softly.

"I know," Nora answered. "So do I, but this is Da and Mama's dream. Da will find work and send for us. Look at the **photographs** that Uncle Sean sent. Doesn't America look grand?"

119

The story and characters are made-up, but the events could happen in real life. Events in historical fiction happened a long time ago.

Illustrations show details about how people lived.

Your Turn

COLLABORATE

Find two things in the story that could happen in real life. Talk about why "Sailing to America" is historical fiction.

Similes

A simile compares two very different things. It uses the word *like* or *as*. This is a simile. *Her cheeks were like red roses.* Similes are different from everyday language.

Find Text Evidence

In "Sailing to America," I see this sentence, "The waves were as big as mountains." This simile compares the way the waves looked with mountains. That means the waves looked huge and tall.

The waves were as big as mountains.

Your Turn   COLLABORATE

Talk about the meaning of this simile from "Sailing to America."

The air inside the steamship smelled like a dirty sock. page 121

Tristan Elwell

Readers to...

Writers choose words to make their stories clear. They use interesting nouns to name people, places and things. Reread the passage from "Sailing to America."

Precise Nouns

Find the noun *steamship*. What other words mean about the same thing? How does **word choice** help make the story clear?

Expert Model

A year later, Da had saved enough money to send for his family. Mama, Danny, and Nora packed what little they had. They got on a crowded steamship and began their voyage.

The trip across the Atlantic Ocean was rough. The air inside the steamship smelled like a dirty sock. The ship tossed up and down for days. The waves were as big as mountains. Many passengers became seasick, but Nora and Danny felt fine.

Tristan Elwell

Writers

Abby wrote about a family's move to a new home. Read her revisions.

Editing Marks

≡ Make a capital letter.

/ Make a small letter.

⊙ Add a period

∧ Add,

✐ Take out.

Grammar Handbook

Singular and Plural Nouns
See page 479.

Student Model

Ben's New Home

In March, Ben and his ~~mother~~ ^family^

~~and father and sister~~ moved to

Tampa. Ben's father got a new job~~s~~.

At first Ben felt sad⊙ Then he met

new ^s^ friend. He liked his school. Ben

wus happy in his new ^neighborhood^ ~~home~~.

by Abby J.

Your Turn

COLLABORATE

✔ Identify interesting nouns.

✔ Identify singular and plural nouns.

✔ Tell how revisions improved the writing.

Go Digital!
Write online in Writer's Workspace

129

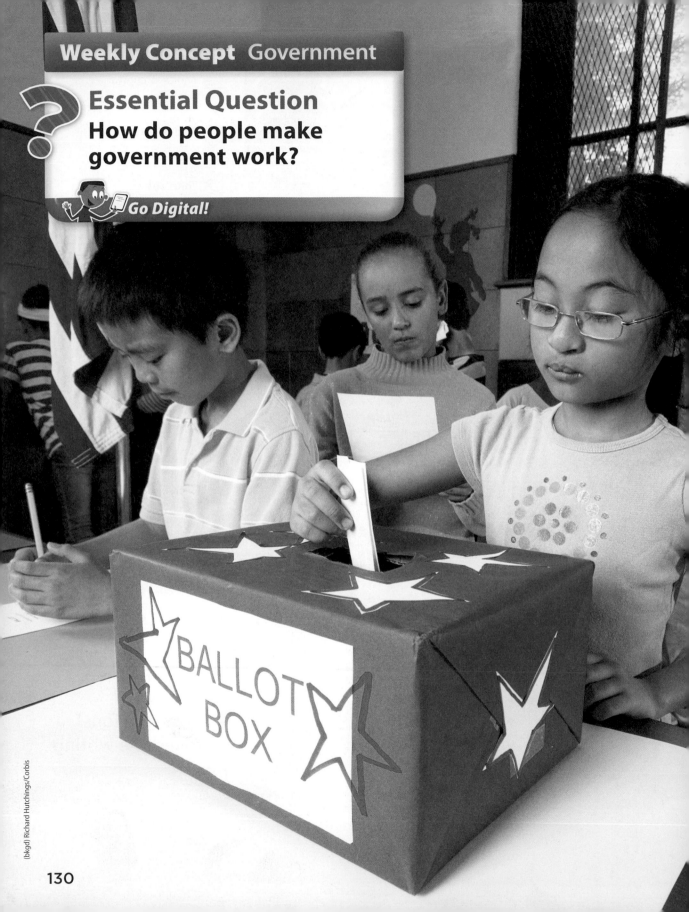

Essential Question
How do people make government work?

Go Digital!

Our Voices Count

Today we are voting for class president. I will vote for Jan. I think she will do the best job. My vote will count!

▶ Voting lets people know what you think.

▶ It gives you the power to choose.

Talk About It

Write words about voting. Talk with a partner about why voting is so important.

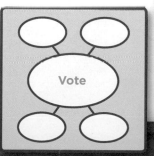

Vocabulary

Use the picture and the sentence to talk with a partner about each word.

announced

Ms. Parks **announced** the winner of the contest.

When has someone announced your name?

candidates

Andrew is one of four **candidates** for class president.

What are some things candidates do before an election?

convince

Julia's mother tried to **convince** her to play soccer.

What is something someone tried to convince you to do?

decisions

Nick made two **decisions** about what to eat for breakfast.

Name two decisions you make every day.

elect

The students voted to **elect** a new class president.

What is another word for elect?

estimate

Sam tried to **estimate**, or guess, the number of coins he had.

What does it mean to estimate something?

government

Our **government** makes laws.

Think of one law our government has made.

independent

It's good to be **independent** and do things on your own.

How can you be more independent at home?

COLLABORATE

Your Turn

Pick three words. Write three questions for your partner to answer.

Go Digital! *Use the online visual glossary*

Every VOTE Counts!

Essential Question

How do people make government work?

Read about one group that teaches kids the power of voting.

Vote for the Class Pet

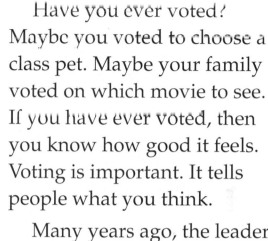

Have you ever voted? Maybe you voted to choose a class pet. Maybe your family voted on which movie to see. If you have ever voted, then you know how good it feels. Voting is important. It tells people what you think.

Many years ago, the leaders of our country wanted to know what people thought, too. They wrote a plan for our **government**. It is called the Constitution. It gives men and women in the United States the right to vote.

Each year, people who are eighteen years and older pick new leaders. They also **vote** on new laws. Voting gives Americans the **power** to choose.

Teaching Kids to Vote

Did you know that only about six out of every ten Americans vote? That's sad. Some people think that voting is too hard. They are unsure where to go to vote. They think it takes too much time. Now, a group called Kids Voting USA is trying to **convince** everyone to vote.

Kids Voting USA teaches kids that voting is important. The group gives teachers lessons to use in their classrooms. First, kids read stories and do fun activities about government. They also learn how to choose and **elect** a good leader.

Election Day is here!

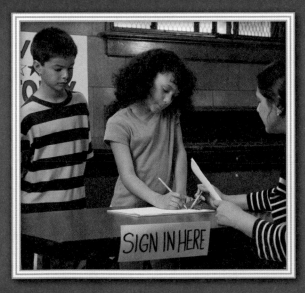

First we sign in.

Richard Hutchings/Corbis

Next, kids talk with their families. They reread stories about **candidates** These are the people who want to be chosen as leaders. Families discuss their ideas and make **decisions**. That way, when it's time to vote, kids know who they want to vote for.

On **election** day, kids get to vote just like adults. They use ballots like the ones in real elections. A ballot is a special form with the names of candidates on it. Kids mark their choices on the ballot. Then they put the ballot into a special box. Finally, all the votes are counted and recounted. The winners are **announced**, and everyone knows who won.

Then we mark a ballot.

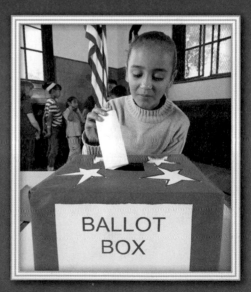

Finally we vote!

Richard Hutchings/Corbis

137

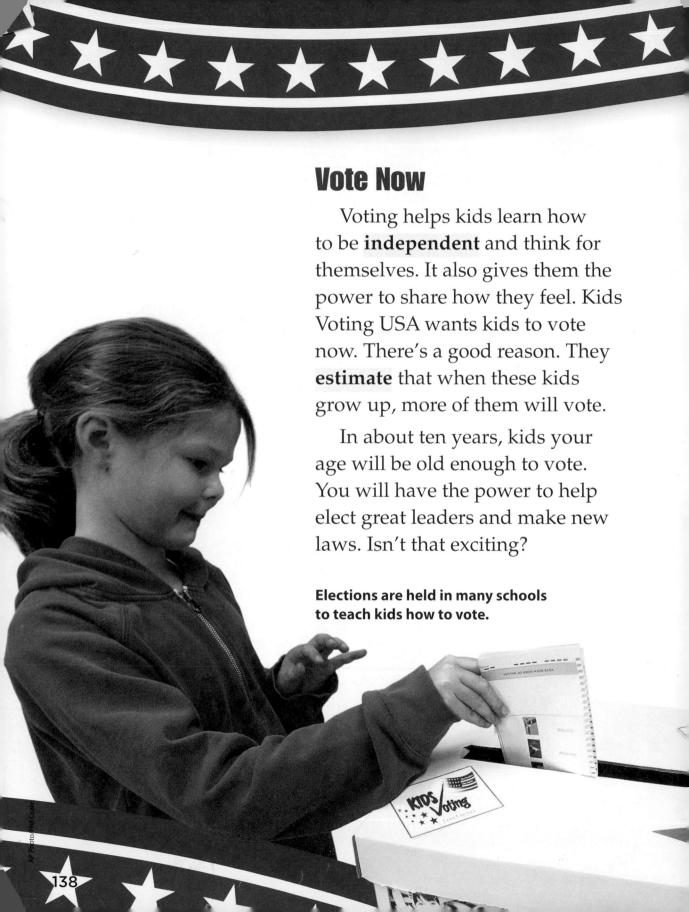

Vote Now

Voting helps kids learn how to be **independent** and think for themselves. It also gives them the power to share how they feel. Kids Voting USA wants kids to vote now. There's a good reason. They **estimate** that when these kids grow up, more of them will vote.

In about ten years, kids your age will be old enough to vote. You will have the power to help elect great leaders and make new laws. Isn't that exciting?

Elections are held in many schools to teach kids how to vote.

This bar graph shows the results of a class election. Which pet was the favorite?

Vote for a Class Pet

Hamster

Hermit Crab

Guinea Pig

Mouse

0 1 2 3 4 5 6 7 8

Make Connections

Talk about voting. How does voting give people the power to choose?
ESSENTIAL QUESTION

Tell about a time when you voted. How did it make you feel? **TEXT TO SELF**

Reread

Stop and think about the text as you read. Do you understand what you are reading? Does it make sense? Reread to make sure you understand.

Find Text Evidence

Do you understand why the author thinks voting is important? Reread the first part of page 135.

page 135

Have you ever voted? Maybe you voted to choose a class pet. Maybe your family voted on which movie to see. If you have ever voted, then you know how good it feels. Voting is important. It tells people what you think.

Many years ago, the leaders of our country wanted to know what people thought, too. They wrote a plan for our **government**. It is called the Constitution. It gives men and women in the United States the right to vote.

Each year, people who are eighteen years and older

I read that <u>voting is a way to tell people what you think. It is a way for people to choose new laws and leaders.</u> Now I understand why the author thinks voting is important.

Your Turn

COLLABORATE

How does Kids Voting USA teach kids to vote? Reread pages 136 and 137.

Author's Point of View

An author often has a point of view about a topic. Look for details that show what the author thinks. Then decide if you agree with the author's point of view.

 Find Text Evidence

What does the author think about voting? I can reread and look for details that tell me what the author thinks. This will help me figure out the author's point of view.

Details
The title of the story is "Every Vote Counts!"
The author thinks it's sad that only six out of every ten Americans vote.
Voting gives Americans the right to choose.

Details help you figure out the author's point of view.

Point of View
Voting is important. Everyone should vote.

Your Turn

Reread "Every Vote Counts!" Find details that tell how the author feels about Kids Voting USA. Write them in a graphic organizer. What is the author's point of view? Do you agree with it?

Go Digital!
Use the interactive graphic organizer

Expository Text

"Every Vote Counts!" is an expository text.

Expository text:
- Gives facts and information about a topic
- Has headings that tell what a section is about
- Includes text features, such as bar graphs

Find Text Evidence

I can tell "Every Vote Counts!" is expository text. It gives facts about voting. It also has headings and a bar graph.

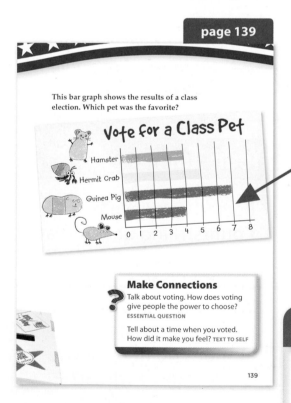

page 139

This bar graph shows the results of a class election. Which pet was the favorite?

Vote for a Class Pet

Hamster
Hermit Crab
Guinea Pig
Mouse

0 1 2 3 4 5 6 7 8

Make Connections

Talk about voting. How does voting give people the power to choose? ESSENTIAL QUESTION

Tell about a time when you voted. How did it make you feel? TEXT TO SELF

139

Text Features

Headings Headings tell what a section of text is mostly about.

Bar Graph A bar graph is a special kind of picture. It helps you understand numbers and information in a quick and easy way.

COLLABORATE

Your Turn

Look at the bar graph on page 139. Tell your partner something you learned from it.

Prefixes

A prefix is a word part added to the beginning of a word. It changes the meaning of the word. The prefix *un-* means "not." The prefix *re-* means "again."

 Find Text Evidence

In "Every Vote Counts!" I see this sentence. "They reread stories about candidates." *The word* reread *has the prefix* re-. *I know that the prefix* re- *means "again." The word* reread *must mean "read again."*

They <u>reread</u> stories about candidates.

Your Turn

Find prefixes. Figure out the meanings of the following words in "Every Vote Counts!"
unsure, *page 136*
recounted, *page 137*

Readers to...

Writers use details to support and explain their ideas. Details help readers understand the topic. Reread the passage from "Every Vote Counts!"

Details

How do voters use ballots on election day? Identify **details** that help readers understand voting.

Expert Model

On **election** day, kids get to vote just like adults. They use ballots like the ones in real elections. A ballot is a special form with the names of candidates on it. Kids mark their choices on the ballot. Then they put the ballot into a special box. Finally, all the votes are counted and recounted. The winners are announced, and everyone knows who won.

BALLOT BOX

Writers

Victor wrote about how he feels about voting. Read Victor's revision.

Editing Marks

☰ Make a capital letter.

／ Make a small letter.

⊙ Add a period.

⌃ Add.

✒ Take out.

Grammar Handbook

Irregular Plural Nouns
See page 479.

Student Model

★ ★ ★ Vote! ★ ★ ★

We live in the United States.

We have the right to vote⊙ Voting is

a very important thing to do. When

people vote, they help choose new

leaders. ~~Vote~~ Voting is like a job. ~~Childs~~ Children

can vote in school. everyone

should vote.

by Victor M.

Ballot
☐ Jen
☐ Nate
☑ Maria

COLLABORATE

Your Turn

- ✔ Find details that explain an idea.
- ✔ Identify an irregular plural noun.
- ✔ Tell how revisions improved the writing.

Go Digital!
Write online in Writer's Workspace

? **Essential Question**
How can people help
animals survive?

Go Digital!

SAVING ANIMALS

These manatees look healthy and happy, but sometimes they need help. We can help threatened animals, like manatees, survive.

► We can protect their habitats.

► We can keep their habitats clean.

► We can join special groups that help animals.

Talk About It

Write words you have learned about what animals need to survive. Talk with a partner about things you can do to save animals.

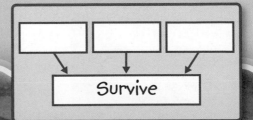

Survive

Vocabulary

Use the picture and the sentence to talk with a partner about each word.

caretakers

The **caretakers** at the zoo feed the penguins every day.

What would caretakers for a horse do?

population

There is a large **population** of flamingos living in the pond.

Name another animal population that might live in a pond.

recognized

Nola **recognized** herself in the mirror.

What is another word for recognized?

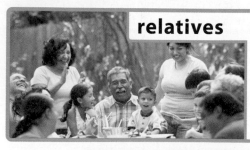

relatives

We invited friends and **relatives** to our picnic.

How do you spend time with your relatives?

resources

Plants need **resources**, such as sunlight and fresh air, to grow.

What resources do people need?

success

Jill's performance was a **success**.

Tell about a success you have had.

survive

Animals need water to **survive**.

What are some other things animals need to survive?

threatened

The wildfire **threatened** the trees.

Describe a time when a storm threatened your plans.

Your Turn

Pick three words. Write three questions for your partner to answer.

Go Digital!   *Use the online visual glossary*

KIDS to the Rescue!

Olivia and Carter Ries,
founders of One More Generation

? Essential Question

**How can people help
animals survive?**

Read how two children helped
sea turtles survive an oil spill.

(Inset) Courtesy of One More Generation

150

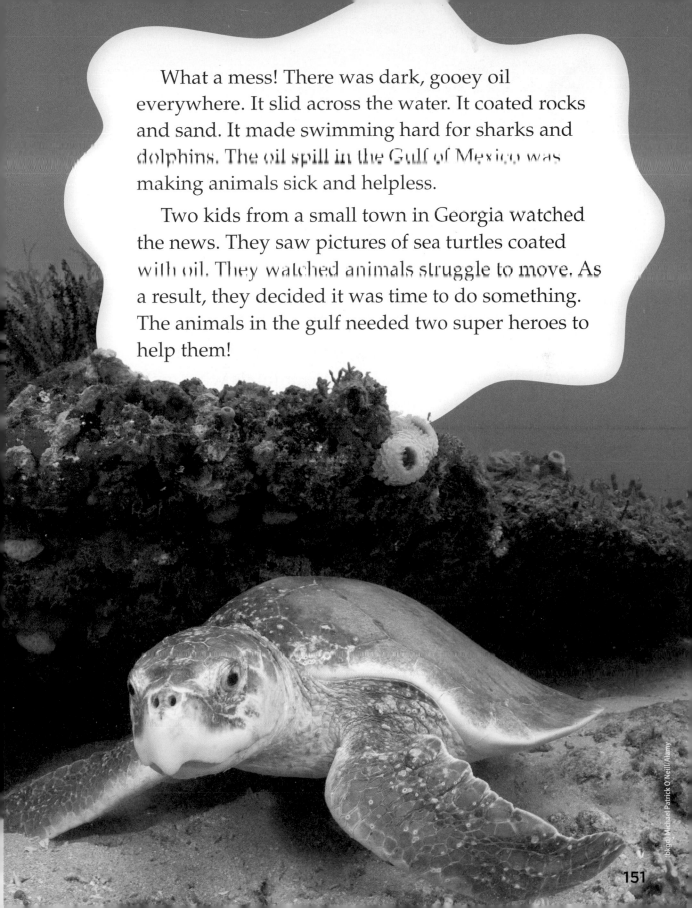

What a mess! There was dark, gooey oil everywhere. It slid across the water. It coated rocks and sand. It made swimming hard for sharks and dolphins. The oil spill in the Gulf of Mexico was making animals sick and helpless.

Two kids from a small town in Georgia watched the news. They saw pictures of sea turtles coated with oil. They watched animals struggle to move. As a result, they decided it was time to do something. The animals in the gulf needed two super heroes to help them!

Olivia and Carter to the Rescue!

Meet Olivia and Carter Ries. They started a group that works to save animals. Olivia was seven years old, and Carter, her brother, was eight-and-a-half. They named their group One More Generation. They want animals to be around for kids in the future.

Olivia and Carter believe everyone can make a difference. They are sending an important message. Their message is that everyone can help animals.

Olivia and Carter watched oil spread for miles across the gulf. More and more animals were getting sick. The Kemp's ridley turtle was one of them. There are only a few thousand left in the world. They are endangered, and their **population** is getting smaller and smaller. The oil **threatened** to ruin their homes and their habitat.

Olivia and Carter Ries learned how oil harms Kemp's ridley turtles.

This female turtle is clean and healthy. It is being returned to the gulf.

Oil Spoils Everything

Olivia and Carter learned that the female turtles were swimming across the gulf to Mexico. They were going to lay eggs on the beaches there. But the thick oil destroyed the **resources** the turtles need to live. The harmful oil covered the sand. It made it hard for them to swim.

Sea turtles **survive** by eating seaweed, jellyfish, and small sea animals. The oil spill spoiled their food, too. Without food, the turtles die.

Saving the Sea Turtles

Olivia and Carter **recognized** how big the problem was. The turtles needed help. First they made a thoughtful plan. Then they called a rescue group in New Orleans. They found out that the workers needed useful cleaning supplies and wipes. Next, the kids asked friends, **relatives**, and people in their town to help. They told them how the donations would help remove oil from the turtles.

Olivia and Carter collected supplies for four months. They rode with their parents to New Orleans. They carried the supplies with them. Then the kids watched **caretakers** clean hundreds of sea turtles. With the help of many people, the turtles were soon spotless. Olivia and Carter's plan worked. It was a huge **success**!

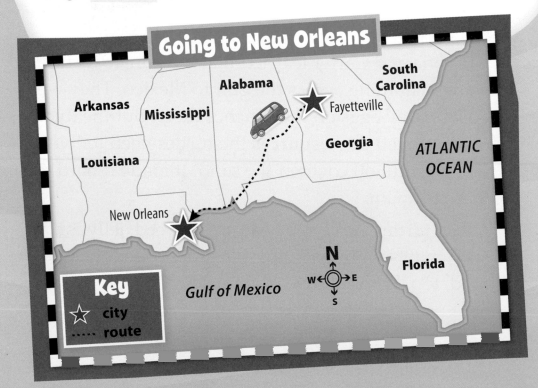

Going to New Orleans

Arkansas

Mississippi

Alabama

South Carolina

Fayetteville

Georgia

ATLANTIC OCEAN

Louisiana

New Orleans

N
W   E
S

Florida

Key
☆ city
····· route

Gulf of Mexico

TSI Graphics

Keeping Busy

Olivia and Carter work with many other groups to help animals all over the world. They give talks at museums and schools. They ask community leaders to support laws that help animals. They help rescue animals in danger.

Olivia and Carter are truly super heroes to endangered animals. With their help, many animals will survive for one more generation.

Ways You Can Help Animals!

- Protect animal nests.
- Pick up trash at parks and wild places.
- Keep water clean.
- Stop using plastic bags.

Carter and his mom unpack supplies in New Orleans.

Make Connections

Describe the steps that Olivia and Carter took to help the Kemp's ridley sea turtles. **ESSENTIAL QUESTION**

What can you and your friends do to help animals? **TEXT TO SELF**

Reread

Stop and think about the text as you read. Are there new facts and ideas? Do they make sense? Reread to make sure you understand.

 Find Text Evidence

Do you understand why an oil spill is harmful to animals? Go back and reread page 151.

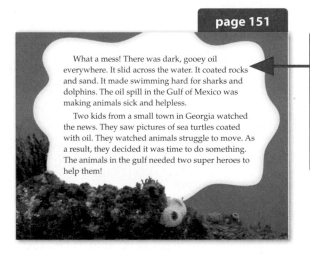

page 151

What a mess! There was dark, gooey oil everywhere. It slid across the water. It coated rocks and sand. It made swimming hard for sharks and dolphins. The oil spill in the Gulf of Mexico was making animals sick and helpless.

Two kids from a small town in Georgia watched the news. They saw pictures of sea turtles coated with oil. They watched animals struggle to move. As a result, they decided it was time to do something. The animals in the gulf needed two super heroes to help them!

I read that oil made it hard for sharks and dolphins to swim. Sea turtles were coated with oil. They struggled to move. These details help me understand why oil is harmful to animals.

Your Turn

Reread the section "Olivia and Carter to the Rescue!" Look for details about how oil spills harm animals.

Author's Point of View

An author often has a point of view about a topic. Look for details that show what the author thinks. Then decide if you agree with the author's point of view.

 ## Find Text Evidence

What does the author think about Olivia and Carter's work with animals? I can reread and look for details that tell me what the author thinks.

Details
Olivia and Carter recognize that turtles need help.
Olivia and Carter collected supplies to help the turtles.
Their plan was a huge success.

Point of View

Details from the text help you figure out the point of view.

Your Turn

Reread "Kids to the Rescue." Use the details in your graphic organizer. Write the author's point of view about Olivia and Carter. Do you agree with the author's point of view?

Go Digital!
Use the interactive graphic organizer

Expository Text

"Kids to the Rescue!" is an expository text.

Expository text:
- Gives facts and information about a topic
- Has headings and sidebars
- Includes text features, such as maps

Find Text Evidence

I can tell that "Kids to the Rescue!" is nonfiction. It gives facts and information about a group that helps animals. It also has a sidebar and a map.

page 154

Saving the Sea Turtles

Olivia and Carter **recognized** how big the problem was. The turtles needed help. First they made a thoughtful plan. Then they called a rescue group in New Orleans. They found out that the workers needed useful cleaning supplies and wipes. Next, the kids asked friends, **relatives**, and people in their town to help. They told them how the donations would help remove oil from the turtles.

Olivia and Carter collected supplies for four months. They rode with their parents to New Orleans. They carried the supplies with them. Then the kids watched **caretakers** clean hundreds of sea turtles. With the help of many people, the turtles were soon spotless. Olivia and Carter's plan worked. It was a huge **success**!

Going to New Orleans

Arkansas Mississippi Alabama South Carolina Fayetteville Georgia ATLANTIC OCEAN Louisiana New Orleans Florida

Key
☆ city
- - - route

Gulf of Mexico

154

Text Features

Sidebar A sidebar gives more information about a topic.

Map A map is a picture of an area. It shows cities, roads, and rivers.

COLLABORATE

Your Turn

Look at the text features in "Kids to the Rescue!" Tell your partner something you learned.

Suffixes

A suffix is a word part added to the end of a word. It changes the word's meaning. The suffix -*ful* means "full of." The suffix -*less* means "having no" or "without."

 Find Text Evidence

In "Kids to the Rescue!" I see this sentence, "The harmful oil covered the sand." *The word* harmful *has the suffix* -ful. *I know that the suffix* -ful *means "full of." The word* harmful *must mean "full of harm."*

The _harmful_ oil covered the sand.

Your Turn

Find suffixes. Figure out the meanings of the following words in "Kids to the Rescue!"
helpless, *page 151*
thoughtful, *page 154*
spotless, *page 154*

Readers to...

Writers use the words *first*, *next*, *then*, and *finally* to show time order, or sequence. These words help put events in order. Reread the passage from "Kids to the Rescue!"

Sequence

Find two sequence words. How do they help the author **organize** the events?

Expert Model

Olivia and Carter recognized how big the problem was. The turtles needed help. First they made a thoughtful plan. Then they called a rescue group in New Orleans. They found out that the workers needed useful cleaning supplies and wipes. Next, the kids asked friends, relatives, and people in their town to help. They told them how the donations would help remove oil from the turtles.

One More Generation

Writers

Ryan wrote about helping dogs at an animal shelter. Read his revisions.

Editing Marks

≡ Make a capital letter.

/ Make a small letter.

⊙ Add a period.

∧ Add.

✗ Take out.

Grammar Handbook

Combining Sentences: Nouns
See page 480.

Student Model

Help for Strays

Last week our class help the
                          helped

Greensville Animal Shelter. Here

is what we did. We collected pet
                    and
food, We collected old rags. Next
          carried
we carryed them to the shelter.
                    , and
Then we cleaned cages. We walked

the dogs. Finally, we gave all the

animals some food?

by Ryan L.

Dog

Your Turn

COLLABORATE

- ☑ Identify sequence words.
- ☑ Find combined sentences.
- ☑ Tell how revisions improved the writing.

Go Digital!
Write online in Writer's Workspace

161

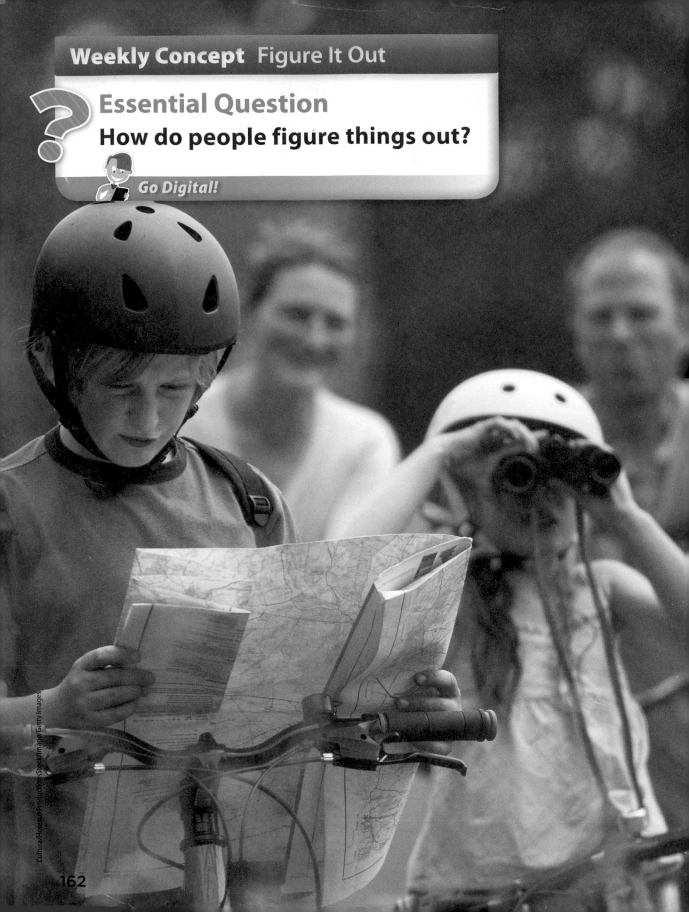

Essential Question

How do people figure things out?

Go Digital!

What Do I Do Next?

When my family and I go biking, it's my job to figure out the best way to go. I use a map. It helps me imagine the best places to ride. Then we all talk about it.

▶ I also ask questions that help me decide what to do.

▶ There are many ways to figure things out.

Talk About It

Think about how you decide what to do. Talk with a partner. Write down how you figure things out.

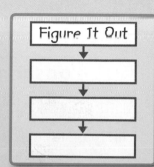

Vocabulary

Use the picture and the sentence to talk with a partner about each word.

bounce

Keith likes to **bounce** a soccer ball off his head.

How many times can you bounce a ball without stopping?

imagine

Mandy likes to **imagine** what her dream house might look like.

What do you imagine when you daydream?

inventor

Thomas Edison was the **inventor** of the first light bulb.

What does an inventor do?

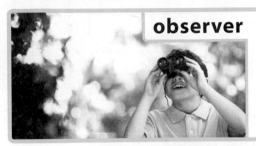

observer

Jason is a good **observer** and enjoys watching birds.

Tell about a time when you were an observer at an event.

Poetry Words

alliteration

"Poets paint precise pictures" is an example of **alliteration**.

Give another example of alliteration.

free verse

Jeremy likes to write **free verse** poems because they don't need to rhyme.

What would you write a free verse poem about?

limerick

Dan's **limerick** had five lines and made the class laugh.

How is a limerick different from other poems?

rhyme

The words *cat* and *bat* **rhyme** because they end in the same sound.

Name two other words that rhyme.

COLLABORATE

Your Turn

Pick three words. Then write three questions for your partner to answer.

Go Digital! **Use the online visual glossary**

Empanada Day

One bite of Abuelita's empanadas
And my mouth purrs like a cat.
　"Teach me," I beg and **bounce** on my feet,
　"Teach me to make this magical treat."
Abuelita smiles,
　"Be an **observer**, watch and learn,
　Then you too can take a turn."

Essential Question

How do people figure things out?

Read poems about different ways
to figure things out.

Dara Goldman

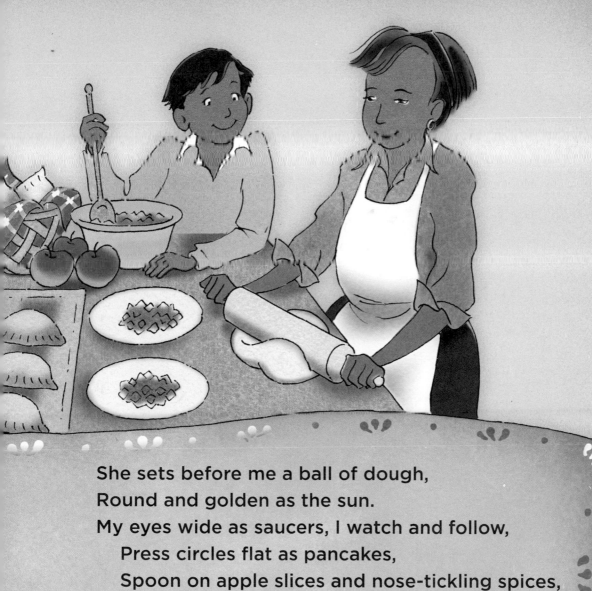

She sets before me a ball of dough,
Round and golden as the sun.
My eyes wide as saucers, I watch and follow,
    Press circles flat as pancakes,
    Spoon on apple slices and nose-tickling spices,
    Seal it all in, a half-moon envelope of bliss.
Together we write down every step
As the empanadas bake and crisp in the oven,
My stomach rumbling like a hungry bear.
                    Ah, empanada day!

                        – George Santiago

Cold Feet

An **inventor** with feet like ice,

And toes like ten shivering mice,

Looked at clothes, studied feet.

Read about cold and heat,

And knit the first socks, warm and nice.

OUR WASHING MACHINE

Our washing machine is a bear

That munches up socks by the pair.

He will suds them and grumble

As they spin, turn, and tumble,

Then spit them out, ready to wear.

Dara Goldman

168

Bugged

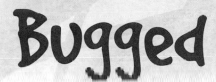

A creature has crawled on my knee,

It's a bug green and round as a pea.

His five wings are fish fins,

He's got teeth sharp as pins.

Just **imagine** him chomping on me!

I read every bug book I see,

To learn what this creature might be.

I ask scientists too,

But they don't have a clue.

So I'm bugged by this great mystery.

Make Connections

What are different ways to figure things out? Talk about what happens in each poem.
ESSENTIAL QUESTION

Which poem has the best way to solve a problem? **TEXT TO SELF**

Limerick and Free Verse

Limerick: • Is a short funny poem that rhymes. • Each stanza has five lines. • The first, second, and fifth lines rhyme. The third and fourth lines rhyme.

Free verse: • Does not always rhyme. • Can have any number of lines and stanzas.

 Find Text Evidence

I can tell that "Cold Feet" is a limerick. It is funny. The stanza has five lines. The first, second, and fifth lines rhyme. The third and fourth lines also rhyme.

page 168

Cold Feet

An inventor with feet like ice,
And toes like ten shivering mice,
Looked at clothes, studied feet.
Read about cold and heat,
And knit the first socks, warm and nice.

OUR WASHING MACHINE

Our washing machine is a bear
That munches up socks by the pair.
He will suds them and grumble
As they spin, turn, and tumble,
Then spit them out, ready to wear.

168

In this funny limerick, the first, second, and fifth lines rhyme. This limerick has one stanza. A stanza is a group of lines in a poem.

Your Turn

COLLABORATE

Reread the poems "Our Washing Machine" and "Empanada Day." Explain whether each poem is free verse or a limerick.

Point of View

A poem often shows a narrator's thoughts about events or characters. This is the point of view. Look for details that show point of view.

 Find Text Evidence

I'll reread "Empanada Day" and look for details that show what the narrator thinks about making empanadas with Abuelita, his grandmother. This is his point of view.

Details
One bite of Abuelita's empanadas and my mouth purrs.
Teach me to make this magical treat.
My eyes wide as saucers, I watch and follow.

↓

Point of View

Your Turn

Reread "Empanada Day." Think about the details. Then write the narrator's point of view in the graphic organizer. Do you agree with the narrator? Why or why not?

Alliteration and Rhyme

Poets use alliteration and rhyme to make descriptions fun to read and poems sound musical.

 Find Text Evidence

Reread "Bugged" on page 169 aloud. Listen for beginning sounds that repeat. Listen for words that rhyme.

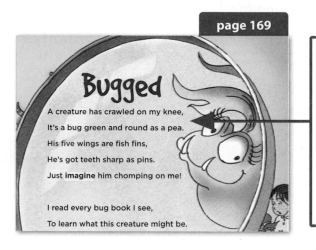

page 169

Bugged

A creature has crawled on my knee,
It's a bug green and round as a pea.
His five wings are fish fins,
He's got teeth sharp as pins.
Just **imagine** him chomping on me!

I read every bug book I see,
To learn what this creature might be.

In the first line, the words crawled *and* creature *start with the same sound.*

The words knee *and* pea *rhyme. I like the way these words sound.*

 COLLABORATE

Your Turn

Find more examples of alliteration and rhyming words in "Bugged."

Dara Goldman

Simile

A simile compares two things that are very different. Similes always have the word *like* or *as*. Two examples of similes are, "The bug is as green as grass" or "The moon is like a giant pearl."

 Find Text Evidence

To find a simile, I need to look for two things that are being compared. In "Cold Feet," I see the line, "An inventor with feet like ice." The simile compares the inventor's feet with ice. That means his feet were very cold.

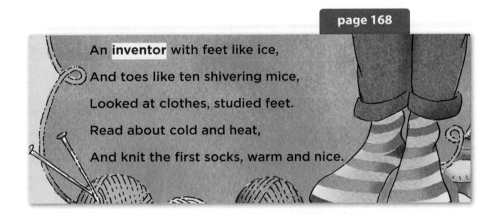

page 168

An **inventor** with feet like ice,
And toes like ten shivering mice,
Looked at clothes, studied feet.
Read about cold and heat,
And knit the first socks, warm and nice.

COLLABORATE

Your Turn

Reread "Bugged." Look for a simile. Talk about the two things being compared.

Dara Goldman

Readers to . . .

Writers use descriptive words and phrases when they write poems. Descriptions help readers create a picture in their minds. Reread "Our Washing Machine" below.

Expert Model

Details

Identify **descriptive details**. How do they help you visualize, or picture, what the washing machine does?

Our washing machine is a bear

That munches up socks by the pair.

He will suds them and grumble

As they spin, turn, and tumble,

Then spit them out, ready to wear.

Dara Goldman

Writers

Abby wrote a description. Read her revisions.

Editing Marks

- ≡ Make a capital letter.
- / Make a small letter.
- ⊙ Add a period.
- ^ Add.
- ↩ Take out.

Grammar Handbook

Possessive Nouns
See page 480.

Student Model

Our Dish Washer

My ~~families~~ family's dish washer is a big cleaning box⊙ It gulps down dirty dish^es^. It sprays and spits water on them. Soon they are as white as pearls. my (≡) ~~mom~~ mom's dishes are always clean and bright.

by Abby K.

Your Turn

- ✔ Identify descriptive details.
- ✔ Identify possessive nouns.
- ✔ Tell how revisions improved her writing.

Go Digital!
Write online in Writer's Workspace

(l) Matti Niemi/Folio Images/Getty Images; (r) Leander Baerenz/The Image Bank/Getty Images

Unit 3

One of a Kind

The Big Idea

Why are individual qualities important?

If Everything

If everything was just the same,
　　　How very dull the world would be.
No different colors, unique sounds,
　　　No mountain, desert, forest, sea.

Imagine everyone alike,
　　　How tiresome they all would seem.
No different sizes, shapes, or thoughts,
　　　And no one left to think and dream.

— **by Winifred Califano**

?

Essential Question
What makes different
animals unique?

Go Digital!

SPECIAL QUALITIES

Bottlenose dolphins are unique mammals. They have the right shape for gliding through the water. And they talk to each other by whistling.

▶ All animals have qualities that are unique.

▶ Animals use their special features to get what they need, protect themselves, and communicate.

Talk About It

Talk with a partner about other animals and their unique qualities. Write words you know here.

Vocabulary

Use the picture and the sentence to talk with a partner about each word.

disbelief

Winnie stared in **disbelief** at the huge shark.

What is another word for disbelief?

dismay

Marco looked at the rain with sadness and **dismay**.

What is the opposite of dismay?

fabulous

The fireworks were amazing and **fabulous**.

What do you think is fabulous?

features

Lions have special **features** that help them survive.

Name a feature that helps lions survive.

offered

Brian **offered** to help Donna get up.

How has someone offered to help you?

splendid

Katie's day at the zoo ended with a wonderful, **splendid** surprise.

List words that mean the same as splendid.

unique

The anteater's long nose makes it a **unique** animal.

What word is the opposite of unique?

watchful

The ducklings learned to swim under the **watchful** eyes of their mother.

What does it mean to be watchful?

Your Turn

COLLABORATE

Pick three words. Write three questions for your partner to answer.

Go Digital! **Use the online visual** ~~gi~~

INCHWORM'S TALE

? Essential Question

What makes different animals unique?

Read about how one animal uses its special features to solve a problem.

Long ago, Anant and his sister, Anika, went swimming. They swam all afternoon and became very tired. They were exhausted and climbed onto a large, flat rock to rest. Soon they fell asleep.

A strange and mysterious thing happened as they slept. The rock beneath them grew and expanded until it reached the clouds.

Anant awoke and looked around. "Sister, wake up!" he cried in **disbelief**. "Am I dreaming, or are we among the clouds?"

Anika rubbed her eyes. "You're not dreaming, brother. This rock has grown while we slept!" The children looked around and saw **fabulous** blue sky and wonderful white clouds.

The children were so high, Anika felt dizzy. Anant searched for a way to climb down, but he could not find a path. Anant and Anika started to cry. They felt fear and **dismay**.

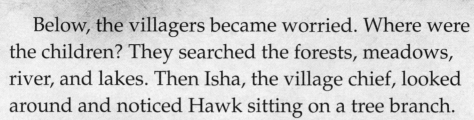

Below, the villagers became worried. Where were the children? They searched the forests, meadows, river, and lakes. Then Isha, the village chief, looked around and noticed Hawk sitting on a tree branch.

"Hawk, will you help us find Anant and Anika?" he asked. "You have sharp, **watchful** eyes and strong wings. They are your best **features**. Please use them to help us find the children."

Hawk agreed to help and flew up into the sky. He tilted his head and squinted his eyes at the bright sunlight. When he was near the clouds, he spied the children on the rock.

"Don't be afraid," said Hawk. "We will rescue you!"

Hawk was unable to carry the children down the rock, so he gathered lots of food for them to eat. Then he brought large leaves to keep them warm. Hawk wanted to make sure they were safe and unharmed.

Hawk flew down to the village and spoke to Isha. Isha called all the animals together and told them they needed help to get the children down. He asked each animal to use its most special feature to climb the tall rock. Several tried and failed.

Mouse's teeth were strong and **unique**, but they couldn't help her climb up the rock.

Bear's huge claws were good for climbing up trees. However, they could not help him scale rocks.

Mountain Lion's claws were sharp and powerful, but the rock was too slippery and he slid back down.

Finally, a tiny voice filled with enthusiasm spoke up and **offered** to help. "May I try, please? It's me, Too-Tock, the Inchworm!"

Inchworm showed them all how skillful she was at climbing. Hawk volunteered to carry Inchworm to the top of the rock. Then she could lead Anant and Anika down the giant mountain. Isha agreed to the plan.

So Hawk carefully picked up Inchworm in his beak. Together they flew to the top of the rock where the children were waiting. Along the way, Inchworm planned for the trip down.

Jago Silver

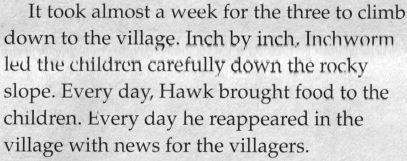

It took almost a week for the three to climb down to the village. Inch by inch, Inchworm led the children carefully down the rocky slope. Every day, Hawk brought food to the children. Every day he reappeared in the village with news for the villagers.

Finally, Inchworm, Anant, and Anika reached the bottom of the rock. Everyone cheered and called Inchworm a hero. It was a glorious, **splendid** day.

"From this day on," said Isha, "I rename the big rock, Too-Tock-Awn-oo-Lah, after the brave inchworm."

Make Connections

What unique feature does Inchworm have? How does it help? **ESSENTIAL QUESTION**

What do your special features help you do? **TEXT TO SELF**

Visualize

Use colorful words and details to help you visualize, or form pictures, in your mind. This will help you understand the characters' actions and feelings.

 ## Find Text Evidence

You may not be sure how Anant and Anika feel when they wake up. Reread page 183. The details will help you visualize how they feel.

> **page 183**
>
> Long ago, Anant and his sister, Anika, went swimming. They swam all afternoon and became very tired. They were exhausted and climbed onto a large, flat rock to rest. Soon they fell asleep.
>
> A strange and mysterious thing happened as they slept. The rock beneath them grew and expanded until it reached the clouds.
>
> Anant awoke and looked around. "Sister, wake up!" he cried in **disbelief**. "Am I dreaming, or are we among the clouds?"
>
> Anika rubbed her eyes. "You're not dreaming, brother. This rock has grown while we slept!" The children looked around and saw **fabulous** blue sky and wonderful white clouds.
>
> The children were so high, Anika felt dizzy. Anant searched for a way to climb down, but he could not find a path. Anant and Anika started to cry. They felt fear and **dismay**.

I read that <u>Anika felt dizzy and Anant searched for a way down. He could not find a way down and they both began to cry.</u> These details help me figure out how scared Anant and Anika felt.

Your Turn

How do the children get back down to the village? Reread and visualize what happens. Then tell what happens.

Problem and Solution

A plot often has a problem and solution. A problem is something that needs to change or be solved. The solution is how the characters fix the problem.

 Find Text Evidence

On page 183, I read that the rock grew and Anant and Anika were stuck on top. This is the problem. I read that the villagers search for the children. Then Isha asks Hawk for help. These are steps to solving the problem. They are in sequence, or time order.

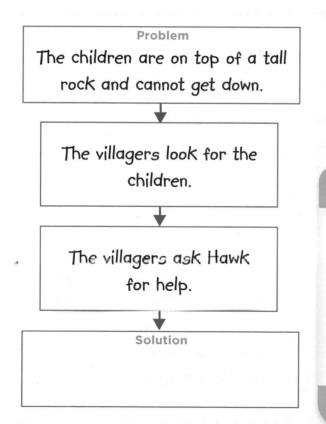

Problem

The children are on top of a tall rock and cannot get down.

↓

The villagers look for the children.

↓

The villagers ask Hawk for help.

↓

Solution

Your Turn

COLLABORATE

Reread "Inchworm's Tale." Find more steps to solving the problem. List them in your graphic organizer in sequence. Then write the solution.

Go Digital!
Use the interactive graphic organizer

Folktale

"Inchworm's Tale" is a folktale.

A **folktale**:

- Is a short story passed from parents to children in a culture
- Usually has a message or teaches a lesson

 Find Text Evidence

I can tell that "Inchworm's Tale" is a folktale. I learned about how a great rock got its name. I also learned a lesson. Even small animals can solve big problems.

page 187

It took almost a week for the three to climb down to the village. Inch by inch, Inchworm led the children carefully down the rocky slope. Every day, Hawk brought food to the children. Every day he reappeared in the village with news for the villagers.

Finally, Inchworm, Anant, and Anika reached the bottom of the rock. Everyone cheered and called Inchworm a hero. It was a glorious, **splendid** day.

"From this day on," said Isha, "I rename the big rock, Too-Tock-Awn-oo-Lah, after the brave inchworm."

Make Connections
What unique feature does Inchworm have? How does it help? ESSENTIAL QUESTION

What do your special features help you do? TEXT TO SELF

187

The lesson in a folktale is sometimes found at the end of the story. It is a message the author thinks is important.

 COLLABORATE

Your Turn

What details in the story show you "Inchworm's Tale" is a folktale? Tell your partner the details.

Synonyms

Synonyms are words that have the same meaning. Sometimes synonyms are context clues for words you don't know.

 Find Text Evidence

On page 183, I'm not sure what expanded *means. I see the context clue* grew *in the same sentence. I know that* grew *means "to get bigger." I think* grew *and* expanded *are synonyms. They have almost the same meanings. Now I know that* expanded *means "grew bigger."*

The rock beneath them grew and expanded until it reached the clouds.

 COLLABORATE

Your Turn

Find synonyms for these words from "Inchworm's Tale."
unharmed, *page 184*
scale, *page 185*

Jago Silver

191

Readers to . . .

Writers use statements, questions, and exclamations. Different kinds of sentences make a story more interesting to read and understand. Reread this passage from "Inchworm's Tale."

Expert Model

Sentence Types

Identify two different **sentence types**. How do they make the story more interesting to read?

Anant awoke, and looked around. "Sister, wake up!" he cried in disbelief. "Am I dreaming, or are we among the clouds?"

Anika rubbed her eyes. "You're not dreaming, brother. This rock has grown while we slept!" The children looked around and saw fabulous blue sky and wonderful white clouds.

The children were so high, Anika felt dizzy. Anant searched for a way to climb down, but he could not find a path. Anant and Anika started to cry. They felt fear and dismay.

Jago Silver

Writers

Marie wrote about her favorite animal. Read Marie's revisions.

Editing Marks

- ≡ Make a capital letter.
- / Make a small letter.
- ⊙ Add a period.
- ∧ Add
- Take out.

Grammar Handbook

Action Verbs
See page 481.

Student Model

Iggy

What swims and is bright orange? ~~.~~ It's my pet fish Iggy! Iggy swims and splashes ~~swims~~ around his tank. his favorite game is to dive to the bottom. Iggy is easy to care for, ~~.~~ and he doesn't make me itch!

Your Turn

- ✔ Identify different types of sentences.
- ✔ Find an action verb.
- ✔ Tell how revisions improved her writing.

Go Digital!
Write online in Writer's Workspace

Essential Question
**How can one person change
the way you think?**

Go Digital!

Jackie Robinson was the first
African-American major league
baseball player.

Bettmann/Corbis

194

MAKE A DIFFERENCE

Jackie Robinson changed the world one home run at a time. He was an inspiring athlete and worked hard to help change the way people treated each other. He was brave and always did his best.

► Brave people lead the way for others.

► They stand up for what they believe in.

► Strong leaders make a difference.

Talk About It

Write words you have learned about leadership. Talk with a partner about what it takes to make a difference.

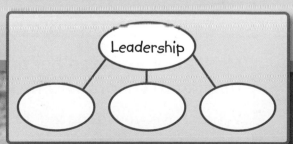

Leadership

John F. Kennedy, U.S. President

Jane Goodall, Scientist

Vocabulary

Use the picture and the sentence to talk with a partner about each word.

amazement

Kris and Lauren watched the movie with **amazement**.

What word means the same as amazement?

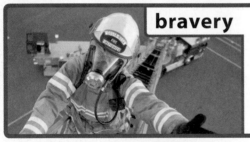

bravery

It takes **bravery** and courage to be a firefighter.

What other jobs take bravery?

disappear

I saw the turtle's head **disappear** into its shell.

What is the opposite of disappear?

donated

Ms. Walker **donated** her time by reading to children at the library.

What have you donated to help others?

leader

Abraham Lincoln was a great **leader**.

Who is another great leader?

nervous

James is **nervous** about speaking in front of his class.

Show how you look when you are nervous.

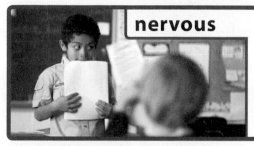

refused

Tom and Kyle **refused** to eat their breakfast.

What is something you refused to do?

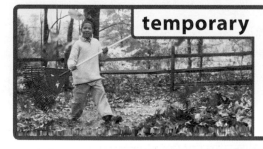

temporary

Raking leaves is a **temporary** job for Steven.

What does it mean when something is temporary?

Your Turn

COLLABORATE

Pick three words. Then write three questions for your partner to answer.

Go Digital! *Use the online visual glossary*

(t) Stacy Gold/National Geographic/Getty Images; (tc) Will & Deni McIntyre/Corbis; (bc) Smith Collection/Taxi/Getty Images; (b) Steve Prezant/Corbis

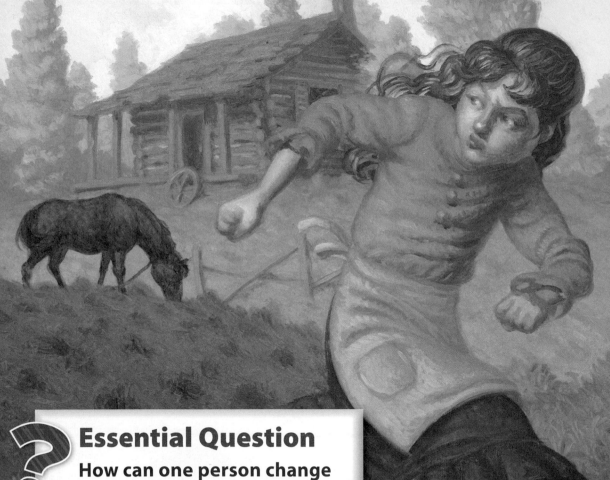

Jane's Discovery

Essential Question

How can one person change the way you think?

Read how a future president changed Jane's life.

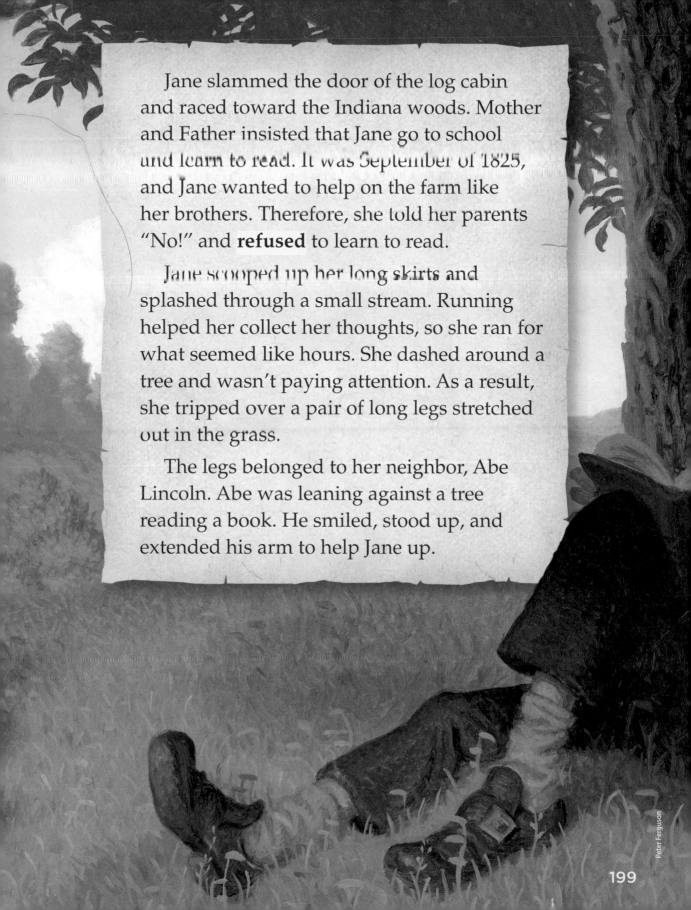

Jane slammed the door of the log cabin and raced toward the Indiana woods. Mother and Father insisted that Jane go to school and learn to read. It was September of 1825, and Jane wanted to help on the farm like her brothers. Therefore, she told her parents "No!" and **refused** to learn to read.

Jane scooped up her long skirts and splashed through a small stream. Running helped her collect her thoughts, so she ran for what seemed like hours. She dashed around a tree and wasn't paying attention. As a result, she tripped over a pair of long legs stretched out in the grass.

The legs belonged to her neighbor, Abe Lincoln. Abe was leaning against a tree reading a book. He smiled, stood up, and extended his arm to help Jane up.

Peter Ferguson

Jane recognized Abe and knew what a hard-worker he was. But she also heard he was not like the other sixteen-year-old boys in Perry County. Abe was different because he spent all of his spare time reading books.

"Why are you running so fast?" Abe asked. "Are you hurt?"

Jane frowned. "No, I'm all right," she said. "I'm running because I'm upset. My parents want me to learn to read, and I told them no!"

Abe looked down at his book and then at Jane.

"Reading can change your life," he said quietly. "Meet me here tomorrow, and I'll prove to you how important reading is."

Jane met Abe the next afternoon. He showed her a book about George Washington. One of his favorite teachers had **donated** it to him, and he had read it many times.

Abe began to read aloud while Jane listened carefully. He read about Washington and what a great **leader** he was. He read about Washington's courage and **bravery**.

"Someday I want to be as courageous as George Washington," said Abe proudly. "Someday I will be president of the United States, too."

"I believe you will make a great president," said Jane. "Look at what a good leader you are now. You've completely changed my mind about reading!"

Abe smiled. "Tell your parents you will learn to read," he said. "Then meet me here every day after school. We will read together and I will help you."

At first, Jane was **nervous** and uncertain about learning to read, but she met Abe every day like clockwork. Fortunately, Jane's intense dislike for reading was only a **temporary** feeling. She was getting the hang of it. As a result, her nervousness began to **disappear**. One afternoon, Abe surprised her. To her **amazement**, he gave her his favorite book.

Peter Ferguson

"Thank you," she said. "Now that I can read, I don't ever want to stop."

Years later, Jane opened her newspaper and read the good news. Her friend, Abe Lincoln, had been elected President of the United States. She smiled and thought about the day she tripped over his long legs. That was the day that changed her life.

PERRY COUNTY TRIBUNE

Lincoln Elected!

Abe Lincoln Elected 16th President

November 6, 1860

Illinois Senator Abraham Lincoln was elected 16th president of the United States of America. He defeated three other candidates in the November 1860 election.

Make Connections

How did Abe change Jane's life? **ESSENTIAL QUESTION**

Who has helped change the way you think? **TEXT TO SELF**

Visualize

Use colorful words and details to help you visualize how characters feel. Picture in your mind what they do as you read.

 Find Text Evidence

How does Jane feel at the beginning of the story? Use the details on page 199 to help you visualize.

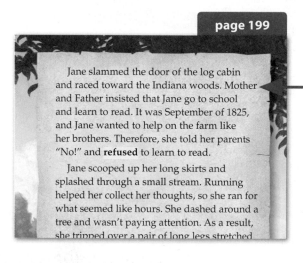

page 199

Jane slammed the door of the log cabin and raced toward the Indiana woods. Mother and Father insisted that Jane go to school and learn to read. It was September of 1825, and Jane wanted to help on the farm like her brothers. Therefore, she told her parents "No!" and **refused** to learn to read.

Jane scooped up her long skirts and splashed through a small stream. Running helped her collect her thoughts, so she ran for what seemed like hours. She dashed around a tree and wasn't paying attention. As a result, she tripped over a pair of long legs stretched

I read that Jane slammed the door and raced into the woods. She ran for a long time. From these details, I can figure out how Jane is feeling. She is upset and angry.

Your Turn

COLLABORATE

What is Abe's first reaction to Jane? Reread and visualize what happens when they meet. Then answer the question.

Peter Ferguson

Cause and Effect

Events In a story's plot are made up of causes and effects. A cause is why something happens. An effect is what happens. Words and phrases, such as *because*, *so*, and *as a result*, often show a cause and its effect.

 Find Text Evidence

On page 199, I read that Jane is upset and angry. That's the effect. Now I need to find the cause. Mother and Father want her to learn to read, and she refuses. That is the cause. A cause and its effect happen in sequence, or time order.

Characters
Jane, Abe

Setting
the woods near Jane's house

Cause	Effect
Jane is upset because Mother and Father want her to learn to read. →	Jane refuses and runs through the woods to calm down.
Cause →	Effect

Your Turn COLLABORATE

What happens next? Reread "Jane's Discovery." Look for more causes and effects. List them in sequence in your graphic organizer.

Go Digital!
Use the interactive graphic organizer

Historical Fiction

"Jane's Discovery" is historical fiction.

Historical fiction:

- Is a story with made-up characters who take part in real events from history
- Has illustrations that show that the story takes place in the past

 Find Text Evidence

I can tell that "Jane's Discovery" is historical fiction. Jane is made up, but she meets and talks with Abraham Lincoln. The story is based on real events that could have happened a long time ago.

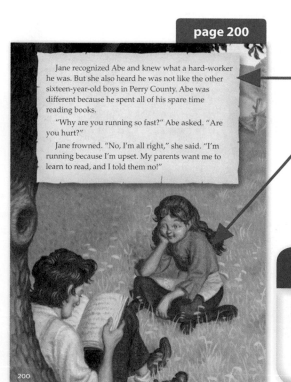

page 200

Jane recognized Abe and knew what a hard-worker he was. But she also heard he was not like the other sixteen-year-old boys in Perry County. Abe was different because he spent all of his spare time reading books.

"Why are you running so fast?" Abe asked. "Are you hurt?"

Jane frowned. "No, I'm all right," she said. "I'm running because I'm upset. My parents want me to learn to read, and I told them no!"

200

In historical fiction, made-up characters might meet with real people from the past.

The illustrations show details about how people dressed.

Your Turn
COLLABORATE

Find two details in "Jane's Discovery" that show this story is historical fiction.

Idioms

An idiom is a group of words that means something different from the meaning of each word in it. The phrase *a piece of cake* is an idiom. It does not mean "a bit of cake." It means "something that's easy to do."

 Find Text Evidence

On page 199, the phrase collect her thoughts *is an idiom. I can use clues in the story to help me figure out that it means "to think about something."*

Running helped her *collect her thoughts,* so she ran for what seemed like hours.

Your Turn

Talk about these idioms from "Jane's Discovery."
like clockwork, *page 202*
getting the hang of it, *page 202*

Peter Ferguson

Readers to...

Writers choose special words to connect, or link, ideas. They use *because, therefore, since, so,* and *for example.* Reread the passage from "Jane's Discovery."

Expert Model

Linking Words

Find a linking word in the sample. Why is this a good **word choice**?

Jane scooped up her long skirts and splashed through a small stream. Running helped her collect her thoughts, so she ran for what seemed like hours. She dashed around a tree and wasn't paying attention. As a result, she tripped over a pair of long legs stretched out in the grass.

Peter Ferguson

Writers

Raj wrote about a time someone inspired him. Read Raj's revisions.

Editing Marks

≡ Make a capital letter.

/ Make a small letter.

⊙ Add a period.

∧ Add.

Take out.

Grammar Handbook

Present Tense Verbs

See page 482.

Student Model

COACH STEVENS

I am a good soccer player. ^because

Coach Stevens helps me. She

inspire~s~ me to do better. ~Last~ ^For example night

she taught us how to pass. She is

very patient. ^and She explains things so

I can understand them.

Your Turn

COLLABORATE

- ☑ Identify linking words.
- ☑ Identify a present tense verb.
- ☑ Tell how revisions improved his writing.

Go Digital!
Write online in Writers' Workspace

209

Essential Question

What do we know about Earth and its neighbors?

Go Digital!

Tom Grundy/Alamy

Discover the Universe

Look up at the sky. What do you see? Astronomers first learned about Earth and its neighbors by looking up.

▶ Today scientists use telescopes, satellites, and manned spaceships to study the universe.

▶ They make new discoveries every day about Earth and our solar system.

Talk About It

Write words you have learned about our solar system. Talk with a partner about these discoveries.

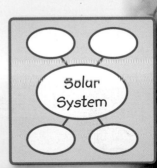

Vocabulary

Use the picture and the sentence to talk with a partner about each word.

amount

James drank a small **amount** of water.

What could you use to carry a large amount of juice?

astronomy

Kia looked at the stars when she studied **astronomy**.

What would you like to learn about astronomy?

globe

Our Earth is a big, round **globe**.

What is another word for globe?

solar system

There are eight planets in our **solar system**.

Name one planet in our solar system.

support

My dad and I **support** our favorite baseball team by cheering.

What can you do to show your support?

surface

An astronaut walks on the dry, dusty **surface** of the moon.

Describe the surface of your desk.

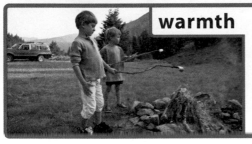

temperature

We can have fun even when the **temperature** outside is cold.

What is the temperature where you are today?

warmth

Will and Paul cooked marshmallows over the **warmth** of a fire.

What is another word for warmth?

COLLABORATE

Your Turn

Pick three words. Then write three questions for your partner to answer.

Go Digital! *Use the online visual glossary*

Earth and Its Neighbors

? Essential Question

What do we know about Earth and its neighbors?

Read about how we have learned about space.

Galileo studied the sky with a telescope he built.

If the Sun could talk, it might say, "Look at me! Look at my sunspots! I am so hot!" Without the Sun, Earth would be a cold, dark planet. How do we know this?

Thanks to the astronomer, Galileo, we know a lot about the Sun and the rest of our **solar system**.

Telescopes: Looking Up

Galileo did not invent the **telescope**. However, 400 years ago he did build one that was strong enough to study the sky. When Galileo looked into space, he saw the rocky surface of the Moon. When he looked at the Sun, he discovered spots on its fiery surface.

The Moon is Earth's closest neighbor.

Astronomy, or the study of space, began with the simple telescope. But astronomers wanted to look at the sky more closely. They made bigger telescopes that could see further than the one Galileo used. Astronomers still had many questions.

Satellites: A Step Closer

In 1958, scientists launched Explorer 1, the first American **satellite**, into space. It was an exciting day for America.

Soon many satellites circled the **globe** and took photographs of Earth, the Moon, stars, and other planets. They collected a large **amount** of information. Satellites even tracked the **temperature** on the planet Saturn.

Explorer 1 takes off.

Scientists have learned many things about the solar system from satellites. That's why they kept sending more into space. Soon there were hundreds of satellites in space making amazing discoveries, but astronomers wanted to know even more. That's why they found a way to put a man on the moon.

(bkgd) NASA, ESA, R. O'Connell (University of Virginia), and the Hubble Heritage Team (STScI/AURA) (c) NASA Marshall Space Flight Center (NASA-MSFC)

One Giant Leap

In 1961, Alan Shepard became America's first **astronaut**. He blasted off into space in a rocket and then turned around and came back to Earth. His short trip was a big success. Shepard's flight proved that people could go into space.

After Shepard, more astronauts went into space. Some orbited the Earth. Some walked on the dusty, bumpy **surface** of the Moon. They took pictures and collected Moon rocks. Astronauts wanted to answer some important questions. Did the Sun's **warmth** heat the moon? Could the Moon **support** life someday?

Astronaut Edwin "Buzz" Aldrin walks toward the Lunar Module. Aldrin left his footprints on the Moon.

Aldrin brought home this Moon rock.

(l) Roger Ressmeyer/Corbis; (r) NASA-JSC

Scientists studied the photographs and Moon rocks that the astronauts brought back. They made exciting discoveries using telescopes and satellites. But it wasn't enough. Scientists wanted to get closer to the other planets. Soon they found a way!

Hubble and Beyond

Scientists created another telescope, but this time it was gigantic. They sent it up into space. The Hubble Space Telescope was launched in 1990. It's still up there and orbits the Earth above the clouds. It takes clear, close-up photographs of stars and planets. It sends fascinating information back to Earth. The Hubble helps scientists study Earth and its neighbors. It also helps astronomers see planets outside our solar system.

It takes the Hubble Telescope 96 minutes to orbit the Earth.

More Discoveries Every Day

Scientists are still asking questions about Earth and its neighbors in space. With the help of satellites, telescopes, and astronauts, they will continue to **explore** and find answers.

What Can We See?

With Our Eyes	With a Simple Telescope	With the Hubble Telescope
The Moon	Craters on the Moon	Planets outside our solar system
The Sun	Sunspots	Stars bigger than the Sun and far, far away
Mars	Clouds around Jupiter	Jupiter's surface

This is a Hubble Telescope photo of an exploding star.

Make Connections

How have we learned about Earth and its neighbors in space? **ESSENTIAL QUESTION**

What do you see when you look at the sky? **TEXT TO SELF**

(b) NASA/ESA/JPL/Arizona State Univ.

Summarize

When you summarize, you tell the most important ideas and details in a text. Use these ideas and details to help you summarize "Earth and Its Neighbors."

 Find Text Evidence

How did telescopes help us learn about space? Identify important ideas and details and summarize them in your own words.

page 215

If the Sun could talk, it might say, "Look at me! Look at my sunspots! I am so hot!" Without the Sun, Earth would be a cold, dark planet. How do we know this?

Thanks to the astronomer, Galileo, we know a lot about the Sun and the rest of our **solar system**.

Telescopes: Looking Up

Galileo did not invent the **telescope**. However, 400 years ago he did build one that was strong enough to study the sky. When Galileo looked into space, he saw the rocky surface of the Moon. When he

I read that Galileo built a telescope. He discovered sunspots and saw the Moon's surface. Details help me summarize. Telescopes helped scientists learn more about space.

Your Turn

COLLABORATE

Reread "Satellites: A Step Closer" on page 216. Summarize the important ideas and details about satellites.

Main Idea and Key Details

The main idea is the most important point an author makes about a topic. Key details tell about the main idea.

 ## Find Text Evidence

I can reread and look for important details about satellites. Then I can figure out what these details have in common to figure out the main idea.

Main Idea
Detail Satellites take photographs of Earth, the Moon, stars, and planets.
Detail
Detail

Your Turn

Reread. Find more key details about satellites. List them in your graphic organizer. Use details to figure out the main idea.

Go Digital!
Use the interactive graphic organizer

221

Expository Text

"Earth and Its Neighbors" is an expository text.

Expository text:
- Gives facts and information about a topic
- Has text features such as headings, key words, and charts

Find Text Evidence

I can tell that "Earth and Its Neighbors" is expository text. It gives facts and information about telescopes, satellites, and space. It has headings, key words, and a chart.

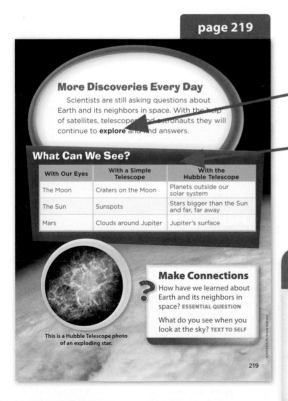

page 219

More Discoveries Every Day

Scientists are still asking questions about Earth and its neighbors in space. With the help of satellites, telescopes and astronauts they will continue to **explore** and find answers.

What Can We See?

With Our Eyes	With a Simple Telescope	With the Hubble Telescope
The Moon	Craters on the Moon	Planets outside our solar system
The Sun	Sunspots	Stars bigger than the Sun and far, far away
Mars	Clouds around Jupiter	Jupiter's surface

This is a Hubble Telescope photo of an exploding star.

Make Connections

How have we learned about Earth and its neighbors in space? ESSENTIAL QUESTION

What do you see when you look at the sky? TEXT TO SELF

219

Text Features

Key Words Key words are important words in the text.

Chart A chart is a list of facts arranged in rows and columns across a page.

COLLABORATE

Your Turn

Look at the chart on page 219. Tell one way the Hubble telescope is different from a simple telescope.

Suffixes

A suffix is a word part added to the end of a word. It changes the word's meaning. The suffix -y means "full of." The suffix -ly means "in a certain way."

 Find Text Evidence

On page 215 in "Earth and Its Neighbors" I see the word rocky. Rocky *has the suffix -y. I know that the suffix -y means "full of." The word* rocky *must mean "full of rocks."*

When Galileo looked into space, he saw the rocky surface of the Moon.

 COLLABORATE

Your Turn

Find the suffix. Use it to figure out each word's meaning.

closely, *page 216*
dusty, *page 217*
bumpy, *page 217*

Readers to . . .

Writers group related ideas together. A strong paragraph has a topic sentence that tells the main idea. Other sentences give details about the main idea. Reread the passage from "Earth and Its Neighbors."

Strong Paragraphs

With a partner, find the topic sentence. Tell how all the ideas in the **paragraph** tell about the main idea.

Expert Model

After Shepard, more astronauts went into space. Some orbited the Earth. Some walked on the dusty, bumpy surface of the Moon. They took pictures and collected Moon rocks. Astronauts wanted to answer some important questions. Did the Sun's warmth heat the moon? Could the Moon support life someday?

NASA-JSC

224

Writers

Steven wrote about the Moon. Read Steven's revisions.

Editing Marks

≡ Make a capital letter.

/ Make a small letter.

⊙ Add a period.

∧ Add.

⌿ Take out.

Grammar Handbook

Past-Tense Verbs
See page 482.

Student Model

The Moon

The Moon is our neighbor in

space. Sometimes it ~~looked~~ looks very

close. The Moon goes around the

Earth⊙ Astronauts have walk ed on

the Moon. They discovered⌿ That

the surface of the Moon is dusty.

There is are lots of

rocks on the Moon.

Your Turn COLLABORATE

- ✔ Identify the main idea.
- ✔ Find a past-tense verb.
- ✔ Tell how revisions improved the writing.

Go Digital!
Write online in Writer's Workspace

Essential Question
What ideas can we get from nature?

Go Digital!

IDEAS FROM NATURE

This spider may be small, but it inspires big ideas. Its webs are super strong, and scientists want to know why.

► Scientists look to nature for new ideas.

► These ideas help people in many different ways.

Talk About It

Talk with a partner about how nature inspires new ideas. Write words you have learned.

Nature

Vocabulary

Use the picture and the sentence to talk with a partner about each word.

effective

The broom is an **effective** tool for sweeping up leaves and dirt.

What is an effective tool for cutting paper?

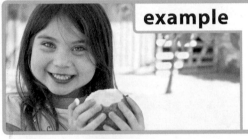

example

The apple is a good **example** of a healthy fruit.

Name an example of a healthy vegetable.

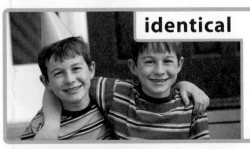

identical

Mark and Matt are **identical** twins because they look alike.

What makes two things identical?

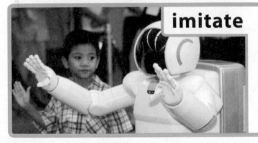

imitate

This robot can **imitate** the way Cody moves.

What does it mean to imitate something?

material

The baby's blanket is made of a soft, warm **material**.

Describe the material your shirt is made of.

model

Kevin and I play with my **model** airplane in the park.

Why does it help to have a model?

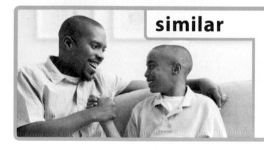

observed

Meg and Joann **observed** the fish, and wrote down what they saw.

What is another word for *observed*?

similar

My dad and I like to look alike, so we wear **similar** shirts.

What is a word that means the opposite of *similar*?

Your Turn

COLLABORATE

Pick three words. Then write three questions for your partner to answer.

Go Digital! **Use the online visual glossary**

BATS DID IT FIRST

Joel Sartore/National Geographic/Getty Images

? Essential Question

What ideas can we get from nature?

Read about how bats inspired a new cane for blind people.

Nature is full of great ideas. Many inventors and scientists just step outside and look around for inspiration and ideas. They often **imitate**, or copy, what they see outdoors. They use nature to inspire their inventions.

One amazing invention was inspired by bats. It's a special cane that helps blind people navigate and get around.

This boy is blind and uses a special cane to help him get around.

Patrick Somelet/Photononstop/GlowImages

231

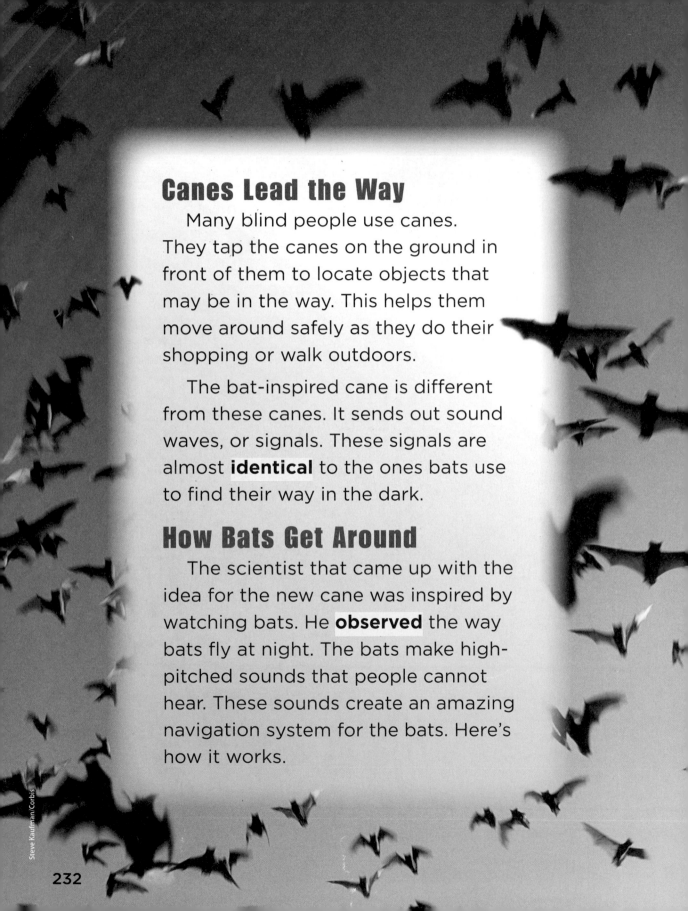

Canes Lead the Way

Many blind people use canes. They tap the canes on the ground in front of them to locate objects that may be in the way. This helps them move around safely as they do their shopping or walk outdoors.

The bat-inspired cane is different from these canes. It sends out sound waves, or signals. These signals are almost **identical** to the ones bats use to find their way in the dark.

How Bats Get Around

The scientist that came up with the idea for the new cane was inspired by watching bats. He **observed** the way bats fly at night. The bats make high-pitched sounds that people cannot hear. These sounds create an amazing navigation system for the bats. Here's how it works.

Bats send sound waves out through their mouth or nose. These sound waves hit objects and then bounce back as an echo. The echo tells the bats how far away an object is and how big it is. This information helps bats find bugs to eat. It is also an **effective** way to keep bats from bumping into trees and other bats.

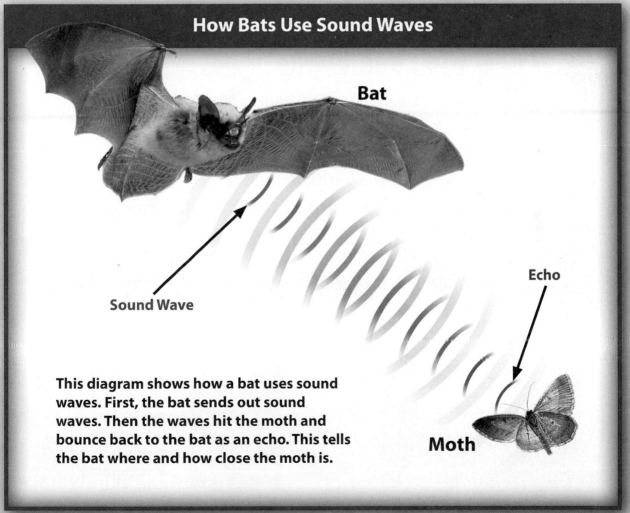

How Bats Use Sound Waves

Bat

Echo

Sound Wave

Moth

This diagram shows how a bat uses sound waves. First, the bat sends out sound waves. Then the waves hit the moth and bounce back to the bat as an echo. This tells the bat where and how close the moth is.

A Batty Idea

The scientist who invented the new cane took what he learned from observing bats. He used a **similar** idea. He started with an ordinary white cane. He wanted the cane to imitate the way bats use sound waves. So, he sketched plans and made a **model** of his invention. When he built the cane, the scientist used a special **material** that was lightweight and strong. Then he added sound waves. Finally, a team of scientists tested the cane. It worked!

How the Cane Works

The handle of the cane sends out signals. The signals bounce off objects in front of the cane. Then an echo bounces back to the cane's handle. The person holding it feels buttons on the handle vibrate, or shake. These buttons tell the person how far away and how big the object is.

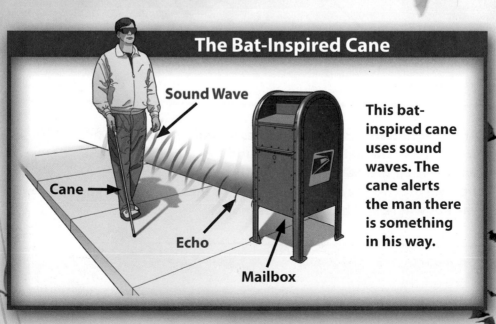

The Bat-Inspired Cane

Sound Wave

Cane

Echo

Mailbox

This bat-inspired cane uses sound waves. The cane alerts the man there is something in his way.

Steve Schell

Scientists and inventors study plants and animals all the time. Their observations have led them to invent many useful things. And like many new inventions, the bat-inspired cane is a good **example** of how great ideas can come from nature.

This scientist is studying how bats fly.

Make Connections

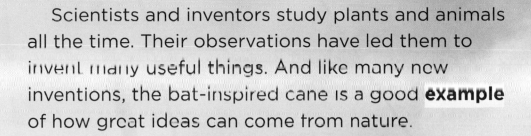

How did bats inspire a cane that helps blind people? ESSENTIAL QUESTION

What is something in nature that inspires you? What would you invent? TEXT TO SELF

Summarize

When you summarize, you tell the most important ideas and details in a text. Use important details to help you summarize "Bats Did It First."

Find Text Evidence

How did one scientist come up with the idea for the new cane? Identify important ideas and details, and summarize them in your own words.

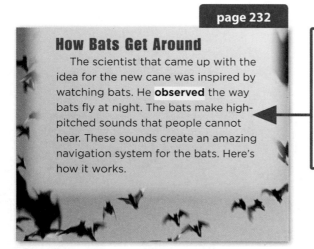

page 232

How Bats Get Around

The scientist that came up with the idea for the new cane was inspired by watching bats. He **observed** the way bats fly at night. The bats make high-pitched sounds that people cannot hear. These sounds create an amazing navigation system for the bats. Here's how it works.

I read that <u>one scientist came up with an idea for a new cane. He watched bats use sound waves to navigate at night.</u> Those details help me summarize. The way bats use sound waves led to the invention of a new cane.

Your Turn

COLLABORATE

Reread "How The Cane Works" on page 234. Summarize the important ideas and details about how the new bat-inspired cane works.

Main Idea and Key Details

The main idea is the most important point the author makes about a topic. Key details tell about the main idea. Put the details together to figure out the main idea.

 Find Text Evidence

What details tell about how bats fly at night? I can reread page 233 and find key details. Then I can figure out what they have in common to tell the main idea.

Main Idea
Detail
Bats make high-pitched sounds through their mouth and nose.
Detail
These sound waves hit objects and bounce back as an echo.
Detail

Your Turn COLLABORATE

Reread. Find more key details about how bats fly at night. List them in your graphic organizer. Then use the details to figure out the main idea.

Go Digital!
Use the interactive graphic organizer

237

Expository Text

"Bats Did it First" is an expository text.

Expository text:
- Gives facts and information about a topic
- Includes text features such as photographs, captions, and a diagram

 Find Text Evidence

I can tell that "Bats Did It First" is an expository text. It has photographs with captions. It also has a diagram that shows how bats fly at night.

page 233

Bats send sound waves out through their mouth or nose. These sound waves hit objects and then bounce back as an echo. The echo tells the bats how far away an object is and how big it is. This information helps bats find bugs to eat. It is also an **effective** way to keep bats from bumping into trees and other bats.

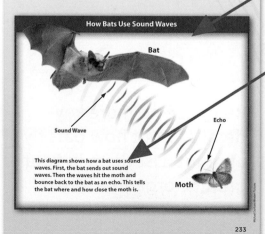

How Bats Use Sound Waves

Bat

Sound Wave

Echo

This diagram shows how a bat uses sound waves. First, the bat sends out sound waves. Then the waves hit the moth and bounce back to the bat as an echo. This tells the bat where and how close the moth is.

Moth

233

Text Features

Diagram A diagram is a picture that gives more information about the text. Labels name the parts of the diagram.

Caption A caption tells about a photograph or diagram.

COLLABORATE

Your Turn

Look at the diagram on page 233. Explain how the bat finds food at night.

Root Words

A root word is the simplest form of a word. It helps you figure out the meaning of a related word.

 Find Text Evidence

In "Bats Did It First" I see the word invention. *I think the root word of* invention *is* invent. *I know* invent *means "to make something new." An* invention *is "something new that is made."*

One amazing invention was inspired by bats.

 COLLABORATE

Your Turn

Find the root word. Then use it to figure out the meaning of each word.

inspiration, *page 231*
navigation, *page 232*

Readers to . . .

A strong conclusion is added at the end of nonfiction writing. It retells the main idea in different words. Reread the passage from "Bats Did It First."

Expert Model

Strong Conclusions

What information is in the **conclusion**?

Scientists and inventors study plants and animals all the time. Their observations have led them to invent many useful things. And like many new inventions, the bat-inspired cane is a good example of how great ideas can come from nature.

Writers

Nick wrote about his favorite idea from nature. Read Nick's revision.

Editing Marks

≡ Make a capital letter.

/ Make a small letter.

⊙ Add a period.

∧ Add

⌇ Take out.

Grammar Handbook

Future-Tense Verbs See page 483.

Student Model

My Favorite Idea from Nature

I want to invent a new kind of
        will
car. I ∧ imitate the tropical boxfish⊙

The car will look a lot like the fish⌇∧

The fish's shape helps it swim

faster underwater. ≡the shape of
                    help
my new car will ~~helped~~∧ it travel

farther on less fuel. ≡i think using

the boxfish to invent a new car is a
    terrific
~~good~~∧ idea.

Your Turn

☑ Identify the strong conclusion.

☑ Identify a future-tense verb.

☑ Tell how revisions improved the writing.

Go Digital!
Write online in Writer's Workspace

241

Essential Question
How is each event in history unique?

Go Digital!

Corbis

History Lives

Christopher Columbus sailed to America in 1492. Today, people reenact his historic voyage. Reliving past events helps us remember them and understand what happened.

► History is made up of many unique events.

► Reliving history helps us appreciate what people went through to get to new places.

Talk About It

Think about a time in history when people moved from one place to another. Talk with a friend about what makes the event unique.

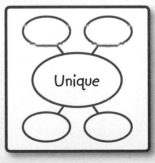

Unique

Vocabulary

Use the picture and the sentence to talk with a partner about each word.

agreeable

Lori loves cold weather and thinks it is pleasant and **agreeable**.

What kind of weather do you find agreeable?

appreciate

Jan and Kayla **appreciate** everything their grandmother does for them.

How do you show people that you appreciate them?

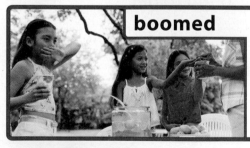

boomed

Anita's lemonade business **boomed** when the weather got hotter.

What does boomed mean in this sentence?

descendants

Ann and her family are **descendants** of the people in the photographs.

What are descendants?

emigration

People who left their homes knew that **emigration** would be hard.

What are some reasons for emigration?

pioneers

In 1843, **pioneers** traveled across the country by covered wagon.

Why did pioneers travel across the country?

transportation

Trains are a favorite form of **transportation** for many people.

Tell about another form of transportation.

vehicles

These **vehicles** are parked in a large parking lot.

What type of vehicle do you travel to school in?

COLLABORATE

Your Turn

Pick three words. Write three questions for your partner to answer.

Go Digital! **Use the online visual glossary**

(t) Archive Holdings/Archive Photos/Getty Images; (tc) Phil Schermeister/National Geographic Society/Getty Images; (bc) Medioimages/Photodisc/Getty Images; (b) Philip and Karen Smith/Iconica/Getty Images

Greg Ryan/Alamy

Essential Question

How is each event in history unique?

Read to see how the pioneers found their way to Oregon.

Pioneers crossed America in covered wagons like these.

The Long Road to Oregon

In the spring of 1843, more than 800 **pioneers** began a journey from Independence, Missouri to Oregon City, Oregon. They had 120 wagons and 5,000 cattle. This was one of the first wagon trains to travel west on the Oregon Trail.

Land of Promise

Life was very hard in Missouri in the 1800s. Bad weather caused crops to die. Businesses closed. As a result, many people could not find jobs.

Americans wanted a more enjoyable life. They wanted to live in a place where the soil was richer and the weather was more **agreeable** and pleasant.

The United States government was giving away free land to pioneers. So, thousands of people traveled west to Oregon, a territory in the northwest part of the country.

This map shows the Oregon Trail in 1843.

The Oregon Trail

Oregon City

OREGON COUNTRY

UNORGANIZED TERRITORY

IOWA

KEY
～ Trail
● Cities

MEXICAN TERRITORY

Independence

MISSOURI

REPUBLIC OF TEXAS

Getting Ready to Go

Pioneers knew that **emigration** to Oregon would be difficult. The trip was more than 2,000 miles of dusty, bumpy trails and would take at least five months. The pioneers needed to be prepared.

First they gathered their cows and chickens. Then they packed many pounds of food, cooking pots, tools, and seeds.

Covered wagons were the main form of **transportation** on the Oregon Trail. The wagons were stuffed with everything a family would need. There was little room for anything else. As a result, children left books, toys, and most of their clothes behind.

A Long, Hard Journey

Planning and packing took weeks. The first thing pioneer families did was hook a team of oxen up to their wagons. Oxen were dependable and strong and could pull the heavy **vehicles**. Next the families joined other pioneer families.

All the wagons traveling together formed a wagon train. The children and healthy adults walked. The sick or tired pioneers rode in the uncomfortable wagons.

Dirty water, sickness, and fierce dust storms made the journey challenging. Bad weather often made the trail impassable. But the pioneers were determined. As a result, they finally reached their new home in Oregon.

This reenactment shows a wagon train on the Oregon Trail.

Greg Ryan/Alamy

A New Life in Oregon

When the pioneers got to Oregon, they cleared land and built houses. Then they planted crops. As more emigrants arrived, towns grew. People opened stores and restaurants. Businesses **boomed**. The pioneers worked hard to make their new towns successful. They had found a better life!

Many of the people who live in Oregon today are **descendants** of the brave pioneers who made the journey west from the 1840s to the 1880s. They **appreciate** their family members' hard work and courage. And they are grateful for the Oregon Trail.

Learn Your History!

History is the study of people and events from the past. It's important to know our country's past. Learning about history helps us appreciate our country and the people who helped build it.

One fun way to learn about history is by reading the stories of the brave people who lived it. You can read diaries of pioneers on the Oregon Trail, or biographies of explorers. These can be more exciting and inspiring than a movie or a television show!

You can still see parts of the original Oregon Trail today.

Make Connections

How was the Oregon Trail emigration a unique time in history? **ESSENTIAL QUESTION**

What is your favorite event in history? Describe why. **TEXT TO SELF**

Summarize

When you summarize, you tell the most important ideas and details in a text. Use details to help you summarize "The Long Road to Oregon."

 Find Text Evidence

Why did pioneers leave Missouri to travel west to Oregon? Reread "Land of Promise" on page 247. Sort the details. Decide which are most important. Then summarize the text in your own words.

---

page 247

In the spring of 1843, more than 800 **pioneers** began a journey from Independence, Missouri to Oregon City, Oregon. They had 120 wagons and 5,000 cattle. This was one of the first wagon trains to travel west on the Oregon Trail.

Land of Promise

Life was very hard in Missouri in the 1800s. Bad weather caused crops to die. Businesses closed.

As a result, many people could not find jobs.

Americans wanted a more enjoyable life. They wanted to live in a place where the soil was richer and the weather was more **agreeable** and pleasant.

The United States government was giving away free land to pioneers. So, thousands of people traveled west to Oregon, a territory in the northwest part of the country.

I read that <u>crops died, businesses closed, and people couldn't find jobs. They got free land. The weather and soil were better in Oregon.</u> These key details help me summarize. Pioneers left Missouri because life was hard and they wanted a better life.

 COLLABORATE

Your Turn

Reread "A New Life in Oregon." Find the key details. Use them to summarize the main ideas and details in your own words.

Sequence

The sequence is the order in which events take place. Look for words and phrases that show time order, such as *first, next, then, later,* and *finally.*

 Find Text Evidence

The events in "The Long Road to Oregon" happen in time order. Life was hard in Missouri in the 1800s, and the United States government was giving away free land. That is the first event. I can use signal words to find more events.

Event

The government was giving away free land in Oregon and the pioneers decided to go west.

↓

First the pioneers gathered their cows and chickens.

↓

Then they packed food, cooking pots, tools, and seeds.

↓

↓

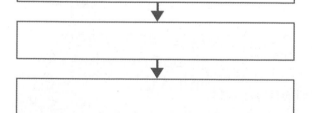

Your Turn

Reread "A New Life in Oregon" on page 249. What happened when the pioneers got to Oregon? List details of the events in order in your graphic organizer. Use signal words to help you.

Go Digital!
Use the interactive graphic organizer

Expository Text

"The Long Road to Oregon" is an expository text.
Expository text:
- May explain a social studies or history topic
- Has headings and sidebars
- May include photos, captions, and maps

 Find Text Evidence

I can tell that "The Long Road to Oregon" is an expository text. It gives information about the Oregon Trail. It also has headings, photographs, captions, and a map.

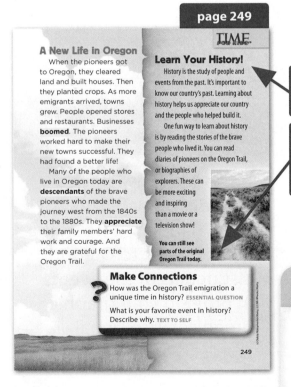

page 249

TIME

A New Life in Oregon
When the pioneers got to Oregon, they cleared land and built houses. Then they planted crops. As more emigrants arrived, towns grew. People opened stores and restaurants. Businesses **boomed**. The pioneers worked hard to make their new towns successful. They had found a better life!

Many of the people who live in Oregon today are **descendants** of the brave pioneers who made the journey west from the 1840s to the 1880s. They **appreciate** their family members' hard work and courage. And they are grateful for the Oregon Trail.

Learn Your History!
History is the study of people and events from the past. It's important to know our country's past. Learning about history helps us appreciate our country and the people who helped build it.

One fun way to learn about history is by reading the stories of the brave people who lived it. You can read diaries of pioneers on the Oregon Trail, or biographies of explorers. These can be more exciting and inspiring than a movie or a television show!

You can still see parts of the original Oregon Trail today.

Make Connections
How was the Oregon Trail emigration a unique time in history? ESSENTIAL QUESTION

What is your favorite event in history? Describe why. TEXT TO SELF

249

Text Features

Sidebar A sidebar may present the author's opinion.

Photographs and captions Photographs and captions give additional facts and details.

Your Turn

 COLLABORATE

Reread the sidebar on page 249. Tell your partner what the author's opinion is.

Suffixes

A suffix is a word part added to the end of a word. It changes the word's meaning. The suffix *able* means "is or can be."

 Find Text Evidence

I see the word enjoyable *on page 247.* Enjoyable *has the root word* enjoy. *I know that* enjoy *means "to be happy with." The suffix -able means "is able or can be." I think the word* enjoyable *means "can be happy with."*

Americans wanted a more enjoyable life.

Your Turn

Use the suffix to figure out the meaning of each word.

agreeable, *page 247*
dependable, *page 248*

Readers to...

Writers use a formal voice for reports. Formal voice has complete sentences and good grammar. Informal voice is less serious and is used when writing for family and friends. Informal voice often uses contractions and slang. Reread the passage from "The Long Road to Oregon."

Formal and Informal Voice

Does the author use a formal or informal voice? How do you know?

Expert Model

In the spring of 1843, more than 800 pioneers began a journey from Independence, Missouri to Oregon City, Oregon. They had 120 wagons and 5,000 cattle. This was one of the first wagon trains to travel west on the Oregon Trail.

Writers

Editing Marks

≡ Make a capital letter.

/ Make a small letter.

⊙ Add a period.

∧ Add.

⟋ Take out.

Shari wrote about why it is important to study history. Read Shari's revision.

Grammar Handbook

Combining Sentences with Verbs See page 485.

Student Model

Study History Now!

Every kid should study history!~~⟋~~

and
~~Every kid should~~ ∧ learn about the

past. ≡Knowing about history makes ⊙

us smart?~~ ∧~~ We can get to know

about important people. ~~We can~~∧ and

learn about their ideas. Studying

history is important.

Your Turn   COLLABORATE

☑ Identify the voice.
☑ Identify combined sentences with verbs.
☑ Tell how revisions improved the writing.

Go Digital!
Write online in Writer's Workspace

255

Unit 4
Meet the Challenge

The Big Idea

What are different ways to meet challenges?

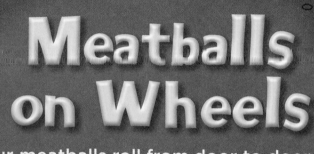

Meatballs on Wheels

Our meatballs roll from door to door,
We go each week at half past four.
We deliver dinner on wheels,
Bringing seniors hot, tasty meals.

Meatballs rolling from door to door,
Meatballs and a whole lot more.

— by Trevor Reynolds

Michael Moran

257

Essential Question

What choices are good for us?

Go Digital!

Smart Choices

Some decisions are easy. This crispy carrot is my favorite snack, and it is delicious and healthy. Making good choices makes me feel good.

▶ We make many decisions every day.

▶ Smart choices help us live healthy lives.

Talk About It

Write words you have learned about choices. Talk with a partner about making smart decisions.

Smart Choices
↓
↓
↓

Vocabulary

Use the picture and the sentence to talk with a partner about each word.

aroma

Carl smells the sweet **aroma** of the flowers near his house.

What is your favorite aroma?

expect

I see clouds, so I **expect** it will rain today.

What do you expect to do while it is raining?

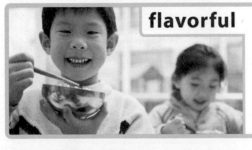

flavorful

Jo and Tori eat lunches that are delicious and **flavorful.**

What are some of your favorite flavorful foods?

graceful

Katie is a **graceful** dancer.

What word means the opposite of graceful?

healthful

Sue chooses **healthful** foods at the market.

What is a good example of a healthful lunch?

interrupted

A small dog **interrupted** the soccer game.

How might you feel if someone interrupted you?

luscious

These strawberries are sweet and **luscious**.

What is another word for luscious?

variety

The bookstore has a large **variety** of books by my favorite author.

Where else could you find a variety of books?

Your Turn

COLLABORATE

Pick three words. Then write three questions for your partner to answer.

Go Digital! *Use the online visual glossary*

Nail Soup

Essential Question

What choices are good for us?

Read about how choices helped a man and his wife learn a lesson.

Illustrator: B Gerardo Suzan

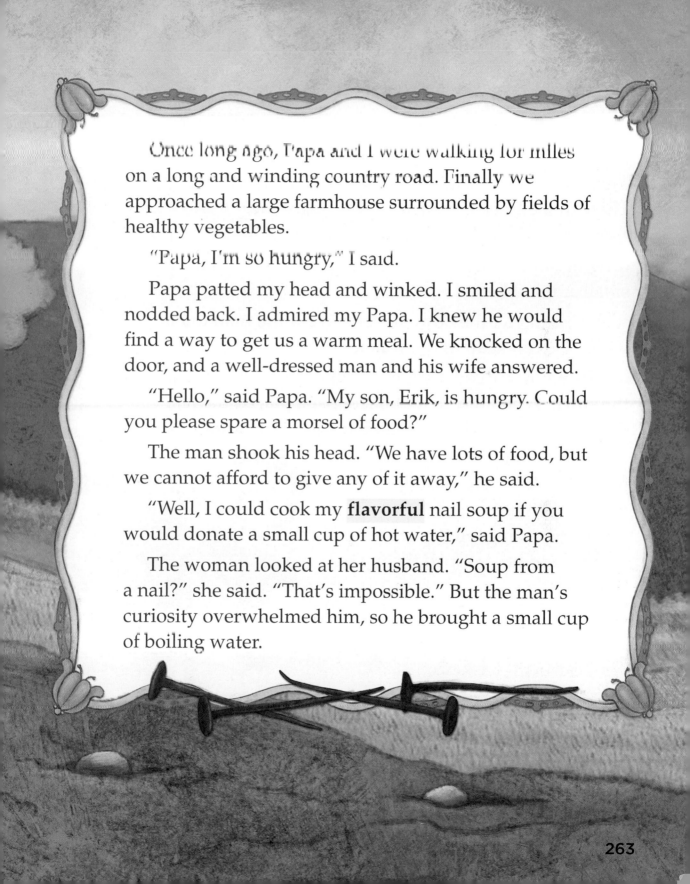

Once long ago, Papa and I were walking for miles on a long and winding country road. Finally we approached a large farmhouse surrounded by fields of healthy vegetables.

"Papa, I'm so hungry," I said.

Papa patted my head and winked. I smiled and nodded back. I admired my Papa. I knew he would find a way to get us a warm meal. We knocked on the door, and a well-dressed man and his wife answered.

"Hello," said Papa. "My son, Erik, is hungry. Could you please spare a morsel of food?"

The man shook his head. "We have lots of food, but we cannot afford to give any of it away," he said.

"Well, I could cook my **flavorful** nail soup if you would donate a small cup of hot water," said Papa.

The woman looked at her husband. "Soup from a nail?" she said. "That's impossible." But the man's curiosity overwhelmed him, so he brought a small cup of boiling water.

Papa carefully took out a long, crooked nail and with one **graceful** motion, dropped it into the cup. He stirred the cup of hot liquid.

"This is beginning to smell wonderful," said Papa.

I smiled at Papa. He was clever and charming, and my admiration for him grew. He could do anything! Then I remembered something he taught me.

"Papa, it is impolite for me to eat nail soup without offering some to everyone," I said. "But there is such a small amount here."

"We can't let the boy eat alone," said the man to his wife. "We can spare more water."

The woman filled a big pot with water and put it on the stove. When the water boiled, Papa placed the nail into the pot, stirred, and sniffed the air. "The **aroma** is good, but it would be much more aromatic with an onion. Have you any old onions?"

The woman gave Papa three small onions, and he dropped them into the pot.

"Papa, remember how **luscious** nail soup was with carrots?" I asked.

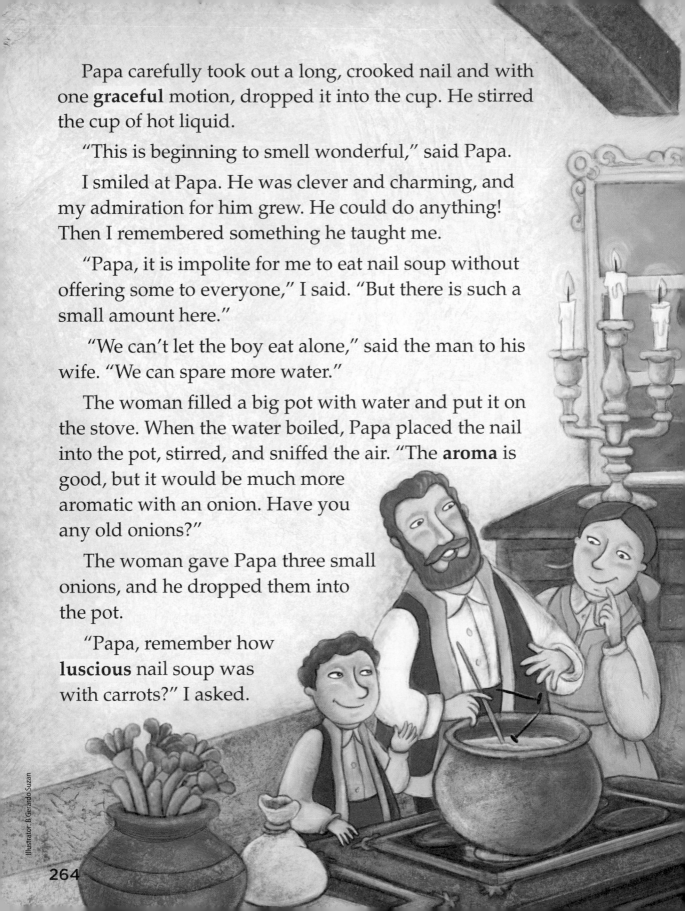

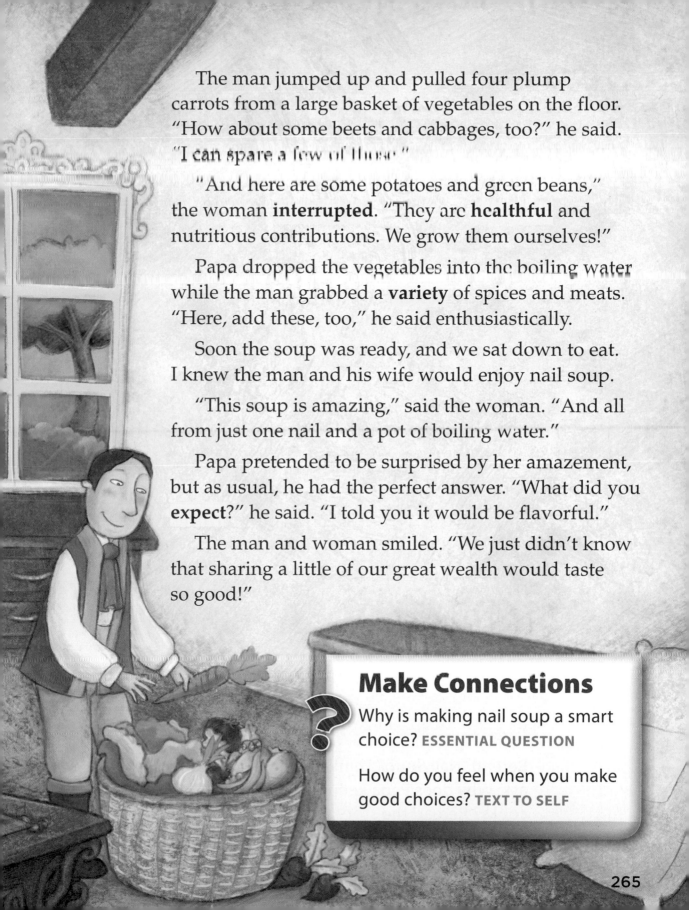

The man jumped up and pulled four plump carrots from a large basket of vegetables on the floor. "How about some beets and cabbages, too?" he said. "I can spare a few of those."

"And here are some potatoes and green beans," the woman **interrupted**. "They are **healthful** and nutritious contributions. We grow them ourselves!"

Papa dropped the vegetables into the boiling water while the man grabbed a **variety** of spices and meats. "Here, add these, too," he said enthusiastically.

Soon the soup was ready, and we sat down to eat. I knew the man and his wife would enjoy nail soup.

"This soup is amazing," said the woman. "And all from just one nail and a pot of boiling water."

Papa pretended to be surprised by her amazement, but as usual, he had the perfect answer. "What did you **expect**?" he said. "I told you it would be flavorful."

The man and woman smiled. "We just didn't know that sharing a little of our great wealth would taste so good!"

Make Connections

Why is making nail soup a smart choice? **ESSENTIAL QUESTION**

How do you feel when you make good choices? **TEXT TO SELF**

265

Ask and Answer Questions

Ask yourself questions about "Nail Soup" as you read. Then look for the details to answer your questions.

Find Text Evidence

Look at page 263. Reread and think of a question. Then read again to answer it.

page 263

Once long ago, Papa and I were walking for miles on a long and winding country road. Finally we approached a large farmhouse surrounded by fields of healthy vegetables.

"Papa, I'm so hungry," I said.

Papa patted my head and winked. I smiled and nodded back. I admired my Papa. I knew he would find a way to get us a warm meal. We knocked on the door, and a well-dressed man and his wife answered.

"Hello," said Papa. "My son, Erik, is hungry. Could you please spare a morsel of food?"

The man shook his head. "We have lots of food, but we cannot afford to give any of it away," he said.

"Well, I could cook my **flavorful** nail soup if you would donate a small cup of hot water," said Papa.

The woman looked at her husband. "Soup from a nail?" she said. "That's impossible." But the man's curiosity overwhelmed him, so he brought a small cup of boiling water.

I have a question. Why did Erik admire Papa? <u>I read that Papa patted Erik's head and winked. He says he knows Papa will find a way to get a warm meal.</u> *Now I can answer my question. Erik admires Papa because he knows Papa will take care of him.*

Your Turn

COLLABORATE

Reread "Nail Soup." Think of a question. You might ask: Why does Papa keep smelling the soup while it cooks? Reread the story to find the answer.

Illustrator: B Gerardo Suzan

Point of View

Point of view is what a narrator thinks about events or other characters in a story. Look for details that show what the narrator thinks to figure out point of view.

 Find Text Evidence

I read on page 263 that Erik, the story's narrator, nodded and smiled at Papa. Then he said he admired his Papa. This tells me he has a lot of respect and love for his father. He trusts he will take care of him.

Details
Erik smiles and nods. He knows Papa will get them a warm meal.

↓

Point of View

 COLLABORATE

Your Turn

Reread "Nail Soup." Look for more clues that show Erik's point of view about Papa. List them in the graphic organizer. Then tell Erik's opinion of his father. Do you agree with Erik's point of view about Papa?

Go Digital!
Use the interactive graphic organizer

Folktale

"Nail Soup" is a folktale. A **folktale**:
- Is a short story passed from person to person
- Always has a problem the characters have to solve
- Usually has a message or lesson

 ## Find Text Evidence

I can tell that "Nail Soup" is a folktale. There is a problem Erik and his Papa have to solve. The story also has a lesson. It teaches the man and his wife the importance of sharing.

page 263

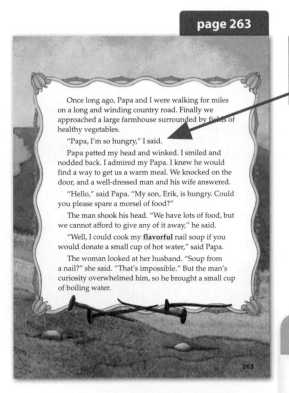

Once long ago, Papa and I were walking for miles on a long and winding country road. Finally we approached a large farmhouse surrounded by fields of healthy vegetables.

"Papa, I'm so hungry," I said.

Papa patted my head and winked. I smiled and nodded back. I admired my Papa. I knew he would find a way to get us a warm meal. We knocked on the door, and a well-dressed man and his wife answered.

"Hello," said Papa. "My son, Erik, is hungry. Could you please spare a morsel of food?"

The man shook his head. "We have lots of food, but we cannot afford to give any of it away," he said.

"Well, I could cook my **flavorful** nail soup if you would donate a small cup of hot water," said Papa.

The woman looked at her husband. "Soup from a nail?" she said. "That's impossible." But the man's curiosity overwhelmed him, so he brought a small cup of boiling water.

263

In this folktale, Erik has a problem. He is hungry. Papa finds a way to solve his problem.

A folktale often has a message that is stated at the end of the story.

Your Turn

COLLABORATE

Reread page 265. What lesson does the man and his wife learn? Tell a partner.

Root Words

A root word is the simplest form of a word. When you read an unfamiliar word, look for a root word in it. Use the root word to figure out the unfamiliar word's meaning.

 Find Text Evidence

On page 264 of "Nail Soup," I see the word admiration. *I think the root word in* admiration *is* admire. *I know* admire *means "to respect or appreciate."* Erik had a lot of admiration *for his* Papa. *That means he "has respect" for him.*

He was clever and charming, and my admiration for him grew.

Your Turn

Find the root word in each word. Use it to figure out the word's meaning.

aromatic, *page 264*
amazement, *page 265*

Illustrator: B Gerardo Suzan

Readers to . . .

Writers use their voice to show how they feel about a character or event in a story. Reread this passage from "Nail Soup."

Show Feelings

How does Erik show his **feelings** for his Papa? Find details that show the reader how Erik feels.

Expert Model

"Papa, I'm so hungry," I said.

Papa patted my head and winked. I smiled and nodded back. I admired my Papa. I knew he would find a way to get us a warm meal. We knocked on the door and a well-dressed man and his wife answered.

"Hello," said Papa. "My son, Erik, is hungry. Could you please spare a morsel of food?"

Illustrator: B Gerardo Suzan

Writers

Ken wrote about why he thinks running is a good choice. Read Ken's revision.

Editing Marks

≡ Make a capital letter.

/ Make a small letter.

⊙ Add a period.

⋀ Add,

𝓎 Take out.

Grammar Handbook

Linking Verbs
See page 481.

Student Model

Running With My Dad

I love running with my dad. We

are running every day after dinner.

think
I ~~thinks~~ running is a good way to

get exercise? I am breathing lots

is
of fresh air. It ~~are~~ so much fun!

We
my dad and I ~~I~~ are a great

running team.

Your Turn

✔ Identify how Ken shows his feelings.
✔ Identify linking verbs.
✔ Tell how revisions improved the writing.

Go Digital!
Write online in Writer's Workspace

Essential Question
How can you use what you know to help others?

Go Digital!

Jeff Greenberg/Alamy

USE YOUR SKILLS

We all have skills and talents. We might be artistic or smart. We might be good at sports or music.

▶ We can use our talents to help others.

▶ Our skills and talents also make us feel good about ourselves.

Talk About It

Write words you have learned about using your talents. Talk with a partner about ways to helps others.

Talents

(t) Corbis

Vocabulary

Use the picture and the sentence to talk with a partner about each word.

achievement

It is a big **achievement** to fly a kite on a very windy day.

What is your biggest achievement?

apologized

Kate **apologized** for breaking the dish.

When have you apologized for doing something?

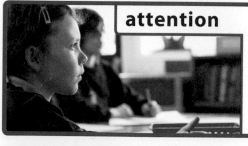

attention

It is important to pay **attention** to directions.

What is something else you should pay attention to?

audience

The **audience** clapped and cheered at the end of the play.

When have you been part of an audience?

confidence

Jody read her report calmly and with **confidence**.

What does it mean to have confidence?

embarrassed

Tia was **embarrassed** when she forgot her lines in the play.

What was something that made you feel embarrassed?

realized

My soccer team celebrated when we **realized** we had won the game.

Describe a time when you realized something.

talents

One of Lila's **talents** is playing the violin.

What talents do you have?

Your Turn

COLLABORATE

Pick three words. Write three questions for your partner to answer.

Go Digital! *Use the online visual glossary*

Andersen Ross/Blend Images/Getty Images; (tc) Hill Street Studios/Blend Images/Getty Images; (bc) John Giustina/Iconica/Getty Images; (b) Bruce Laurance/Blend Images/Getty Images

The Impossible Pet Show

Essential Question

How can you use what you know to help others?

Read how Daniel uses what he knows to save a pet show.

Marcin Piwowarski

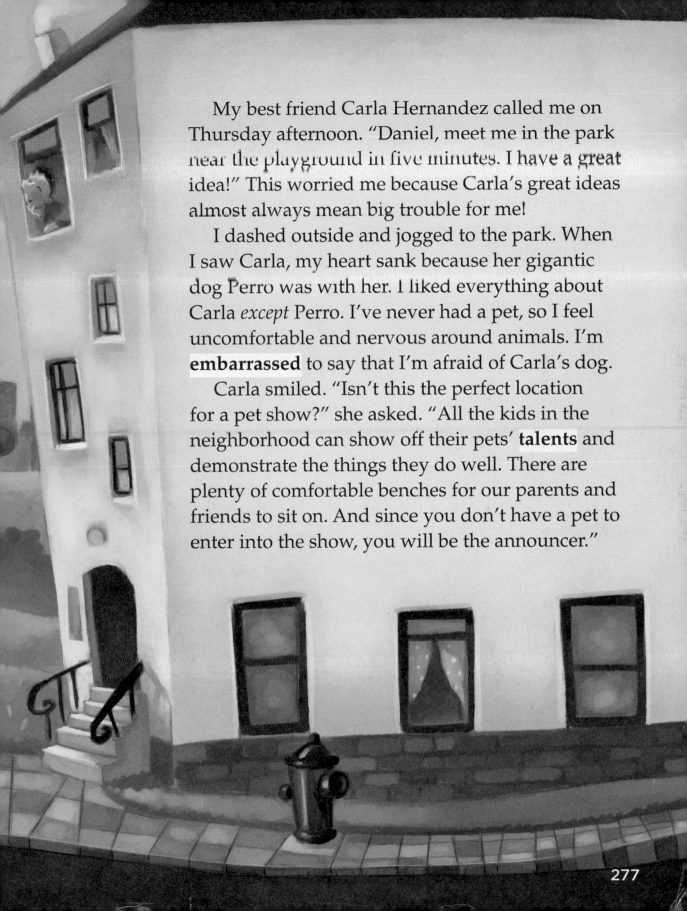

My best friend Carla Hernandez called me on Thursday afternoon. "Daniel, meet me in the park near the playground in five minutes. I have a great idea!" This worried me because Carla's great ideas almost always mean big trouble for me!

I dashed outside and jogged to the park. When I saw Carla, my heart sank because her gigantic dog Perro was with her. I liked everything about Carla *except* Perro. I've never had a pet, so I feel uncomfortable and nervous around animals. I'm **embarrassed** to say that I'm afraid of Carla's dog.

Carla smiled. "Isn't this the perfect location for a pet show?" she asked. "All the kids in the neighborhood can show off their pets' **talents** and demonstrate the things they do well. There are plenty of comfortable benches for our parents and friends to sit on. And since you don't have a pet to enter into the show, you will be the announcer."

"I'm sorry," I **apologized**, "but that's impossible! Crowds make me nervous and unsure. Besides, I don't like animals, remember?"

"That's nonsense," said Carla. "There's nothing to be concerned about because you'll be great!"

Just then, Perro leaped up, slobbered all over me, and almost knocked me down. "Yuck. Down, Perro! Stay!" I shouted. Perro sat as still as a statue. "Wow, you're good at that," said Carla. "Now let's get started because we have a lot to do."

By Saturday morning I had practiced announcing each pet's act a hundred times. My stomach was doing flip flops by the time the **audience** arrived. The size of the crowd made me feel even more anxious.

When the show began, I gulped and announced the first pet. It was a parakeet named Butter whose talent was walking back and forth on a wire. When Butter finished, everyone clapped and cheered. So far, everything was perfect, and I was beginning to feel calmer and more relaxed. I **realized** that being an announcer wasn't so bad after all.

Then it was Carla and Perro's turn.

"Sit, Perro," she said, but Perro didn't sit.

Perro was not paying **attention** to Carla. He was too interested in watching Jack's bunnies jump in and out of their boxes. Suddenly, Perro leaped at the bunnies who hopped toward Mandy and knocked over her hamster's cage. Pudgy, the hamster, escaped and began running around in circles while Kyle's dog, Jake, howled. This was a disaster, and I had to do something.

"Sit!" I shouted at Perro. "Quiet!" I ordered Jake. "Stay!" I yelled. Everyone – kids and pets – stopped and stared at me. Even the audience froze.

"Daniel, that was incredible," said Carla. "You got the pets to settle down. That's quite an **achievement**."

Sadly, that was the end of our pet show. But now I have more **confidence** when I have to speak in front of people. And even though I am still nervous around animals, Perro and I have become great friends. And I've discovered my talent, too.

Make Connections

How did Daniel use what he knows to help others? **ESSENTIAL QUESTION**

Discuss whether you would like to take part in a pet show, and why. **TEXT TO SELF**

279

Ask and Answer Questions

Stop and ask yourself questions about "The Impossible Pet Show" as you read. Then look for story details to answer your questions.

🔍 Find Text Evidence

Reread page 277. Ask a question about what is happening. Then read again to find the answer.

page 277

I dashed outside and jogged to the park. When I saw Carla, my heart sank because her gigantic dog Perro was with her. I liked everything about Carla *except* Perro. I've never had a pet, so I feel uncomfortable and nervous around animals. I'm **embarrassed** to say that I'm afraid of Carla's dog.

Carla smiled. "Isn't this the perfect location for a pet show?" she asked. "All the kids in the neighborhood can show off their pets' **talents** and demonstrate the things they do well. There are plenty of comfortable benches for our parents and friends to sit on. And since you don't have a pet to enter into the show, you will be the announcer."

I have a question. Why are Carla's ideas trouble for Daniel? Daniel is uncomfortable around pets. Carla asks him to help at the pet show. Carla's ideas are trouble because she is asking Daniel to do something he is not comfortable doing.

Your Turn

Reread "The Impossible Pet Show." Think of a question. You might ask: Why does Daniel think being an announcer isn't so bad? Reread page 278 to find the answer.

Point of View

Point of view is what a narrator thinks about other characters or events in a story. Look for details that show what the narrator thinks. Use them to figure out the point of view.

 Find Text Evidence

I read on page 277 that animals make Daniel nervous and uncomfortable. This will help me figure out what Daniel's point of view is about being an announcer for the pet show.

Details
Daniel says he is uncomfortable and nervous around animals.

↓

Point of View

Your Turn

Reread "The Impossible Pet Show." Find more details that tell what Daniel thinks about being an announcer. List them in the graphic organizer. What is his point of view? Do you agree with Daniel's point of view about being an announcer at a pet show?

Go Digital!
Use the interactive graphic organizer

Realistic Fiction

"The Impossible Pet Show" is realistic fiction.

Realistic fiction:

- Is a made-up story that could really happen
- Has dialogue and illustrations
- May be part of a longer book with chapters or part of a series about the same characters

 Find Text Evidence

I can tell that "The Impossible Pet Show" is realistic fiction. The characters talk and act like real people. The events are made up, but they could really happen.

page 278

"I'm sorry," I **apologized**, "but that's impossible! Crowds make me nervous and unsure. Besides, I don't like animals, remember?"

"That's nonsense," said Carla. "There's nothing to be concerned about because you'll be great!"

Just then, Perro leaped up, slobbered all over me, and almost knocked me down. "Yuck. Down, Perro! Stay!" I shouted. Perro sat as still as a statue. "Wow, you're good at that," said Carla. "Now let's get started because we have a lot to do."

By Saturday morning I had practiced announcing each pet's act a hundred times. My stomach was doing flip flops by the time the **audience** arrived. The size of the crowd made me feel even more anxious.

When the show began, I gulped and announced the first pet. It was a parakeet named Butter whose talent was walking back and forth on a wire. When Butter finished, everyone clapped and cheered. So far, everything was perfect, and I was beginning to feel calmer and more relaxed. I **realized** that being an announcer wasn't so bad after all.

Dialogue Dialogue is what the characters say to each other.

Illustrations Illustrations give more information or details about the characters and setting in the story.

Your Turn

Reread "The Impossible Pet Show." Find two events that help you figure out this is realistic fiction.

Prefixes

A prefix is a word part added to the beginning of a word. A prefix changes the word's meaning. The prefixes *un-*, *non-*, and *im-* mean "not" or "opposite of." The prefix *pre-* means "before."

 Find Text Evidence

On page 278, I see the word unsure. *It has the root word* sure *and the prefix* un-. *I know that* sure *means "certain" and the prefix* un- *means "not." The word* unsure *must mean "not certain."*

Crowds make me nervous and unsure.

 COLLABORATE

Your Turn

Use the prefixes in each word to figure out its meaning.

uncomfortable, *page 277*
impossible, *page 278*
nonsense, *page 278*

Marcin Piwowarski

Readers to...

Writers use dialogue and description to show the character's thoughts, feelings, and actions. Reread the passage from "The Impossible Pet Show."

Expert Model

Characters

Read the **dialogue** and **description**. How does the author use them to show what characters are like?

"I'm sorry," I apologized, "but that's impossible. Crowds make me nervous and unsure. Besides, I don't like animals, remember?"

"That's nonsense," said Carla. "There's nothing to be concerned about because you'll be great!"

Just then, Perro leaped up, slobbered all over me and almost knocked me down. "Yuck. Down, Perro! Stay!" I shouted.

Writers

Isabel wrote about how she and her sister solved a problem. Read her revisions.

Editing Marks

≡ Make a capital letter.

/ Make a small letter.

⊙ Add a period.

∧ Add

◞ Take out.

Grammar Handbook

Contractions with *Not* See page 485.

Student Model

The Picnic Problem

My sister s̲u̲e̲ and I went to the park for a picnic. t̲h̲e̲n̲ it started to rain. I jumped up.

"Oh, no!" I sobbed. "This means we can ∧'t ~~not~~ eat our lunch here."

Sue gave me a hug and ∧smiled ~~smile~~.

"Don't worry," she said. "We can ∧have ~~has~~ our picnic at home."

Your Turn

COLLABORATE

☑ Identify dialogue.
☑ Find a contraction.
☑ Tell how revisions improved the writing.

Go Digital!
Write online in Writer's Workspace

285

Essential Question
How do animals adapt to challenges in their habitat?

Go Digital!

Greg Winston/National Geographic/Getty Images

Adapt to Challenges

This ermine's fur is brown and white in the summer. It turns white in the winter and blends in with the snowy ground. This adaptation helps ermines escape its predators.

► Ermines are also fast runners and good climbers.

► They have an excellent sense of smell.

► Adaptations help ermines survive.

Talk About It

Write words you have learned about adaptation. Talk with a partner about ways animals have adapted.

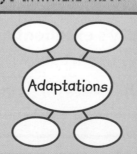

Adaptations

(r) Arco Images GmbH/Alamy

Vocabulary

Use the picture and the sentence to talk with a partner about each word.

alert

Wolves howl to **alert** other wolves when danger is nearby.

How would you alert someone to talk quietly?

competition

Joe won the **competition** because he was the fastest runner.

What kind of competition have you participated in?

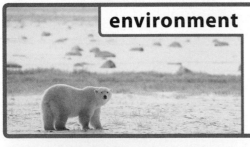

environment

The polar bear lives in a cold and snowy **environment**.

Describe a whale's environment.

excellent

Lily's **excellent** artwork won first place in the art show.

Tell about a time when something you did was excellent.

prefer

Simon and his friends **prefer** walking to riding their bikes.

What kind of transportation do you prefer?

protection

The skunk's scent provides **protection** from its enemies.

Tell what you use for protection on sunny days.

related

Josh and Jen are **related** because they are both members of the same family.

Name two animals that are related.

shelter

Our tent was a dry and safe **shelter** during the storm.

What is another kind of shelter people use?

Your Turn

COLLABORATE

Pick three words. Write three questions for your partner to answer.

Go Digital! *Use the online visual glossary*

GRAY WOLF! RED FOX!

? **Essential Question**

How do animals adapt to challenges in their habitat?

Read how gray wolves and red foxes adapt to challenges.

Did you ever see a photograph of a gray wolf or a red fox? Don't they look a lot like dogs? Aren't they fantastic-looking animals? Well, dogs, foxes, and wolves are all **related**. They are all members of the same family. And while gray wolves and red foxes may look alike, they are different in many ways.

LOOKS ARE EVERYTHING

The gray wolf is the largest member, or a part, of the wild dog family. An adult wolf is the size of a large dog. The red fox is smaller and weighs less. Both animals have **excellent** hearing. The red fox can even hear small animals digging holes underground.

And just take a look at those beautiful tails! The gray wolf and red fox both have long, bushy tails. The wolf's tail can be two feet long. The fox's tail is not as long but has a bright, white tip. In the winter, foxes use their thick, furry tails as **protection** from the cold.

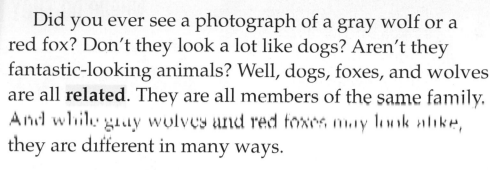

The gray wolf and red fox are both mammals.

Foxes and wolves also have thick fur. Their coats can be white, brown, or black. However, red foxes most often have red fur, while a gray wolf's fur is usually more gray and brown.

FINDING FOOD

Gray wolves and red foxes live in many different habitats. They live in forests, deserts, woodlands, and grasslands. But as more people build roads and shopping centers, both animals have lost their homes. The red fox has adapted well, or made changes, to fit into its **environment**. Now more foxes make their homes close to towns and parks. Wolves, however, stay far away from towns and people.

Foxes and wolves are not in **competition** for food. They have different diets. Red foxes **prefer** to hunt alone and eat small animals, birds, and fish. They also like to raid garbage cans and campsites for food. Wolves work together in packs, or groups, to hunt large animals, such as moose and deer.

WHERE DO THEY LIVE?

United States of America

N
W   E
S

LEGEND
Red Fox only
Gray Wolf only
Both

Gray wolves prefer to live and hunt in packs.

DAY-TO-DAY

Wolves live in packs of four to seven. They do almost everything together. They hunt, travel, and choose safe places to set up dens for **shelter**. Foxes, on the other hand, like to live alone. They usually sleep in the open or find an empty rabbit hole to call home.

The red fox hunts for food alone.

Both wolves and foxes communicate by barking and growling. The gray wolf also howls to **alert**, or warn, other wolves when there is danger nearby. The red fox signals in a different way. It waves its tail in the air to caution other foxes.

The gray wolf and red fox are members of the same family and have many things in common. But they really are two very different animals.

Make Connections

How have the gray wolf and the red fox adapted to living in North America? **ESSENTIAL QUESTION**

Which animal would you like to learn more about? Why? **TEXT TO SELF**

Reread

Stop and think about the text as you read. Are there new facts and ideas? Do they make sense? Reread to make sure you understand.

 Find Text Evidence

Do you understand how red foxes look different from gray wolves? Reread "Looks Are Everything" on page 291.

---

page 291

Did you ever see a photograph of a gray wolf or a red fox? Don't they look a lot like dogs? Aren't they fantastic-looking animals? Well, dogs, foxes, and wolves are all **related**. They are all members of the same family. And while gray wolves and red foxes may look alike, they are different in many ways.

LOOKS ARE EVERYTHING

The gray wolf is the largest member, or a part, of the wild dog family. An adult wolf is the size of a large dog. The red fox is smaller and weighs less. Both animals have **excellent** hearing. The red fox can even hear small animals digging holes underground.

And just take a look at those beautiful tails! The gray wolf and red fox both have long, bushy tails. The wolf's tail can be two feet long. The fox's tail is not as long but has a bright, white tip. In the winter, foxes use their thick, furry tails as **protection** from the cold.

I read that gray wolves are bigger than red foxes. I also read that the color of their fur and their tails look different. Now I understand some of the ways the red fox and gray wolf look different.

Your Turn

 COLLABORATE

Reread the section "Looks Are Everything." Look for details about how gray wolves and red foxes are alike.

Darrell Gulin/Stone/Getty Images; (twigs) McGraw-Hill Companies, Inc.

Compare and Contrast

When authors compare, they show how two things are alike. When they contrast, they tell how two things are different. Authors use signal words such as *both*, *alike*, *same*, or *different* to compare and contrast.

 Find Text Evidence

How are red foxes and gray wolves alike and different? I will reread "Gray Wolf! Red Fox!" and look for signal words.

Wolves	Both	Foxes
The wolf's tail can be two feet long	Thick fur and long, bushy tails.	A fox's tail has a bright, white tip at the end.

COLLABORATE

Your Turn

Reread "Gray Wolf! Red Fox!" Find details that tell how red foxes and gray wolves are alike and different. Add these details to your graphic organizer. What signal words helped you?

Go Digital!
Use the interactive graphic organizer

Expository Text

"Gray Wolf! Red Fox!" is an expository text.

Expository text:
- Gives facts and information to explain a topic
- May be about science topics
- Includes text features such as a map, photographs, and captions

🔍 Find Text Evidence

I can tell that "Gray Wolf! Red Fox!" is expository text. It explains how gray wolves and red foxes are alike and different. It includes a map, photographs, and captions.

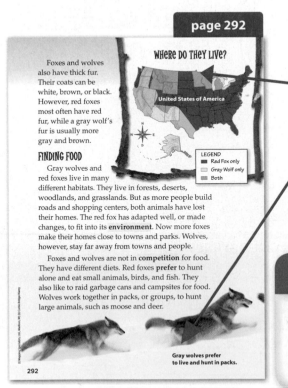

page 292

Text Features

Map A map is a flat drawing of a place. It has a key that shows what colors and symbols mean.

Caption A caption explains a photograph or illustration.

COLLABORATE

Your Turn

Look at the text features in "Gray Wolf! Red Fox!" Tell your partner about something you learned.

Sentence Clues

Sentence clues are words or phrases in a sentence that help you figure out the meaning of an unfamiliar word. Sometimes clues define, or tell exactly, what a word means.

Find Text Evidence

I'm not sure what the word member *means on page 291. I see the words "a part of" in the same sentence. This clue tells me that* member *means "a part of something."*

The gray wolf is the largest member, or a part, of the wild dog family.

Your Turn

Find context clues to figure out the meanings of these words.

adapted, *page 292*
packs, *page 292*

Talk about the sentence clues that helped you figure out the meanings.

Readers to...

A strong opening begins with an interesting question or fascinating fact. It states the topic and grabs the reader's attention. It makes readers want to read more.

Expert Model

Strong Opening

Read the **opening**. Why do the first few lines make you want to read more?

Did you ever see a photograph of a gray wolf or a red fox? Don't they look a lot like dogs? Aren't they fantastic-looking animals? Well, dogs, foxes, and wolves are all related. They are all members of the same family. And while gray wolves and red foxes may look alike, they are different in many ways.

Writers

Nadia wrote about how her favorite animal adapts. Read Nadia's revision.

Student Model

The Biggest Lizards on Earth

Did you know that there are real dragons?
⋀ A ~~Komodo dragon is a lizard.~~

                              are
Komodo dragons ~~is~~ the biggest

lizards on Earth. They are the best

hunters. these dragons have very

sharp teeth? Komodo dragons are

                         am
also very good swimmers. I going to

the zoo to see one today.

Your Turn

COLLABORATE

- ✔ Identify the strong opening.
- ✔ Identify main and helping verbs.
- ✔ Tell how revisions improved the writing.

Go Digital!
Write online in Writer's Workspace

Essential Question
How are people able to fly?

Go Digital!

UP, UP, AND AWAY

People have wanted to fly for hundreds of years. Thanks to many inventors, there are lots of ways they can!

▶ People can cross the country in tiny planes, huge passenger jets, or colorful hot air balloons.

▶ Helicopters take small groups from here to there.

▶ And some adventurers fly solo high above the Earth.

Talk About It

Write words about flight. Talk about how people have learned to fly.

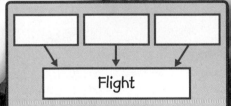

Flight

Vocabulary

Use the picture and the sentence to talk with a partner about each word.

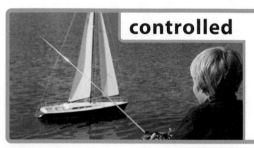

controlled

Tom **controlled** his toy boat's movements from the shore.

What is something you have controlled at home?

direction

The sign showed us which **direction** to go.

Point in the direction of the door.

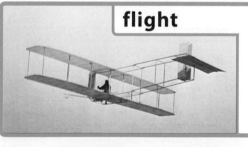

flight

The first airplane **flight** took place many years ago.

Where would you like to go on an airplane flight?

impossible

Crossing this river is **impossible**, so we will have to go a different way.

Name something that is impossible to do.

launched

The space shuttle was **launched** and soared toward space.

What other things can be launched?

motion

Julie enjoys the **motion** of the swing.

What kinds of motion do you like?

passenger

Denise likes being a **passenger** in the car.

When was the last time you were a passenger?

popular

Soccer is the most **popular** sport at our school.

Tell about a popular sport at your school.

COLLABORATE

Your Turn

Pick three words. Write three questions for your partner to answer.

Go Digital! **Use the online visual glossary**

Firsts in Flight

Essential Question

How are people able to fly?

Read about how inventors learned how to fly.

On December 17, 1903, the *Wright Flyer* flew for 12 seconds at Kitty Hawk.

Orville and Wilbur Wright stood on a cold, windy beach in Kitty Hawk, North Carolina. The brothers traveled a long way from their home in Dayton, Ohio to test their newest flying machine. Flying had been their dream since their father had given them a toy helicopter.

Orville and Wilbur Wright

The Wright brothers owned a bicycle shop in Dayton. In addition to selling, building, and repairing bicycles, they built flying machines. They flew the first one in 1899. However, the winds weren't strong enough to keep the machine in **motion**. So they looked for a place where the winds were stronger. As a result, they chose Kitty Hawk. It was not only windy there, but the sandy beaches made for soft landings.

Because their first **flight** was not successful, the Wright brothers learned a lot about flying. As a result, they built a better glider with bigger wings in 1900. This glider did not work very well either. The brothers did not give up. That's why they experimented with a new glider in 1902. Then in 1903, they built the *Wright Flyer*, their first airplane with an engine.

Flying Firsts

By December 17, the brothers were ready to test the *Wright Flyer*. Orville started up the engines to power the plane. He **controlled** the plane, while Wilbur watched from the ground. The *Flyer* was **launched** into the sky. The plane moved in an upward **direction**, and the flight lasted twelve seconds. The Wright brothers had conquered gravity and unlocked the secrets of flying.

Orville and Wilbur kept improving their planes, and their flights became longer. Soon, other people tried to fly airplanes.

Alberto Santos-Dumont was the third man in the world to fly a plane with an engine.

Will It Fly?

Do an experiment on flying using paper airplanes.

Materials needed:

• pencil • paper • ruler

Directions:

1. With a partner, fold two paper airplanes. Make the wing sizes different in each plane.

2. Gently throw one plane.

3. Measure and record how far the paper plane flew.

4. Take turns throwing the plane four more times. Each time, measure and record how far it flies.

5. Repeat the experiment with the other airplane.

6. Compare the plane's flights. Then discuss what you learned about flight.

Heritage Images/Corbis

Alberto Santos-Dumont was an inventor and pilot from Brazil. In 1906, he made the first official flight in front of an audience. The next year, the French pilot, Henri Farman, took along a **passenger** in his plane. They flew for one minute and fourteen seconds.

Better Flying Machines

Because of these flights, airplane research became **popular** with inventors. Before long, better planes were traveling longer distances. In 1909, a French pilot flew an airplane across the English Channel. This plane was very different from the Wright brothers' plane. The new plane had only one long wing across its body. It looked a lot like today's airplanes.

This is what an airplane looked like in 1930.

Soon inventors began building airplanes that could carry more people. By 1920, several new companies offered passengers the chance to fly. Humans had done the **impossible**. They had figured out how to fly.

Make Connections

How did the Wright brothers help people fly? **ESSENTIAL QUESTION**

Tell what you know about airplanes. Discuss other ways to fly. **TEXT TO SELF**

Reread

Stop and think as you read. Does the text make sense? Reread to make sure you understand.

 Find Text Evidence

Do you understand what the Wright brothers learned from their unsuccessful flights? Reread page 306.

---

page 306

Because their first **flight** was not successful, the Wright brothers learned a lot about flying. As a result, they built a better glider with bigger wings in 1900. This glider did not work very well either. The brothers did not give up. That's why they experimented with a new glider in 1902. Then in 1903, they built the *Wright Flyer*, their first airplane with an engine.

Flying Firsts

By December 17, the brothers were ready to test the *Wright Flyer*. Orville started up the engines to power the plane. He **controlled** the plane, while Wilbur watched from the ground. The *Flyer* was **launched** into the sky. The plane moved in an upward **direction**, and the flight lasted twelve

I read that the Wright brothers' first flight was not successful. But they learned a lot about flying. Then they built a better glider with bigger wings. Now I understand why their unsuccessful flights were important.

Your Turn

COLLABORATE

How did other inventors use the Wright brothers' ideas? Reread pages 306 and 307.

Cause and Effect

A cause is why something happens. An effect is what happens. They happen in time order. Signal words, such as *so, as a result, and because* help you find causes and effects.

 ## Find Text Evidence

On page 305 I read that the Wrights had to find a windier place to fly. This is the effect. Now I can find the cause. The wind wasn't strong enough. The signal word so helped me find the cause and effect.

Cause		Effect
First The winds weren't strong enough.	➡	So the brothers found a place where the winds were stronger.
Next	➡	
Then	➡	
Finally	➡	

Your Turn

Reread "Firsts in Flight." Use signal words to help you find more causes and effects. Make sure they are in time order. Fill in the graphic organizer.

Go Digital!
Use the interactive graphic organizer

Expository Text

"Firsts in Flight" is an expository text. **Expository text:**

- May present causes and their effects in sequence
- May explain a science topic
- Includes text features such as headings, photographs, or sidebars

Find Text Evidence

I can tell that "Firsts in Flight" is an expository text. It gives facts and information about how people first started flying. It includes headings, photographs with captions, and a sidebar.

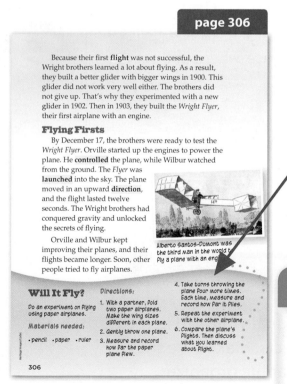

page 306

Because their first **flight** was not successful, the Wright brothers learned a lot about flying. As a result, they built a better glider with bigger wings in 1900. This glider did not work very well either. The brothers did not give up. That's why they experimented with a new glider in 1902. Then in 1903, they built the *Wright Flyer*, their first airplane with an engine.

Flying Firsts

By December 17, the brothers were ready to test the *Wright Flyer*. Orville started up the engines to power the plane. He **controlled** the plane, while Wilbur watched from the ground. The *Flyer* was **launched** into the sky. The plane moved in an upward **direction**, and the flight lasted twelve seconds. The Wright brothers had conquered gravity and unlocked the secrets of flying.

Orville and Wilbur kept improving their planes, and their flights became longer. Soon, other people tried to fly airplanes.

Alberto Santos-Dumont was the third man in the world to fly a plane with an engine.

Will It Fly?

Do an experiment on flying using paper airplanes.

Materials needed:

• pencil • paper • ruler

Directions:

1. With a partner, fold two paper airplanes. Make the wing sizes different in each plane.
2. Gently throw one plane.
3. Measure and record how far the paper plane flew.
4. Take turns throwing the plane four more times. Each time, measure and record how far it flies.
5. Repeat the experiment with the other airplane.
6. Compare the plane's flights. Then discuss what you learned about flight.

306

Text Features

Sidebar A sidebar gives more information about a topic. Sometimes a sidebar can be a science experiment or directions showing how to do something.

COLLABORATE

Your Turn

Look at the text features in "Firsts in Flight." Tell your partner something you learned.

Multiple-Meaning Words

Multiple meaning words have more than one meaning. Find other words in the sentence to help you figure out the correct meaning of a multiple-meaning word.

 Find Text Evidence

On page 306, I know well *can mean "a deep hole with water in it" or "in a good way." The context clue* work *helps me figure out what* well *means in this sentence. I think* well *means "in a good way." The glider did not work in a good way.*

This glider did not work very well either.

Your Turn COLLABORATE

Find context clues. Use them to figure out the meaning of each word.

seconds, *page 306*

fly, *page 307*

Readers to ...

Writers use strong conclusions to retell the main idea and summarize important points. A strong conclusion helps the reader understand the author's purpose.

Strong Conclusions

Identify the **conclusion**. How does it restate the main idea?

Expert Model

Soon inventors began building airplanes that could carry more people. By 1920, several new companies offered passengers the chance to fly. Humans had done the impossible. They had figured out how to fly.

Writers

Marcus wrote about his favorite flying machine. Read his revisions.

Editing Marks

≡ Make a capital letter.

/ Make a small letter.

⊙ Add a period.

∧ Add.

⌔ Take out.

Grammar Handbook

Complex Sentences
See page 477.

Student Model

The Best Flying Machine

My favorite flying machine is

the kite. When I ~~is~~ (am) flying my kite,

I feel happy. It's fun to run. ~~It's~~

~~fun to~~ (and) watch it lift off the ground.

I love the way its (long) tail swishes

back and forth in the wind ⊙ That's

why the kite~~s~~ is my favorite

flying machine of all!

Your Turn

COLLABORATE

☑ Identify the strong conclusion.

☑ Identify a complex sentence.

☑ Tell how revisions improved the writing.

Go Digital!
Write online in Writer's Workspace

Essential Question
How can others inspire us?

Go Digital!

Ira Block/National Geographic/Getty Images

You Inspire Me

Danny talked to the firefighters in his neighborhood. He learned all about what they do. Danny thinks they are brave heroes. He wants to be a firefighter, too.

- ▶ People who are courageous and helpful inspire us.
- ▶ When we feel inspired, we want to help others, too.

Talk About It

Talk with a partner about how we are inspired by others. Write words about inspiration.

Inspiration

315

Vocabulary

Use the picture and the sentence to talk with a partner about each word.

adventurous

Whitewater rafting is an **adventurous** and exciting activity.

What would you like to do that is adventurous?

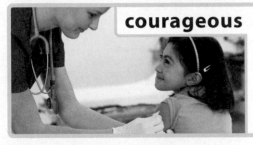

courageous

Maya was **courageous** at the doctor's office.

Describe someone you know who is courageous.

extremely

Some plants can grow in **extremely** dry places.

Tell about a time when you were extremely cold.

weird

The Venus Flytrap is a **weird** and strange plant.

What is a synonym for weird?

Poetry Words

free verse

Some of Emma's **free verse** poems rhyme, and some do not.

Explain why a limerick is not a free verse poem.

narrative poem

I wrote a **narrative poem** about the history of flight.

What kinds of stories make good narrative poems?

repetition

Using the same word several times in a poem is called **repetition**.

Why do poets use repetition?

rhyme

The words *night* and *right* **rhyme** because they end in the same sounds.

Why do poets use words that rhyme?

Your Turn   COLLABORATE

Pick three words. Write three questions for your partner to answer.

Go Digital!   *Use the online visual glossary*

Ginger's Fingers

Diverse Images/Universal Images Group / Getty Images

Ginger's fingers are shooting stars,

They talk of adventurous trips to Mars.

Fingers talking without words,

Signing when sounds can't be heard.

Ginger's fingers are ocean waves,

They talk of fish and deep sea caves.

Fingers talking without words,

Signing when sounds can't be heard.

Ginger's fingers are butterflies,

They talk of a honey-gold sunrise.

Fingers talking without words,

Signing when sounds can't be heard.

? Essential Question

How can others inspire us?

Read about different ways
that people inspire others.

The Giant

Dodge, dart, dash,
    Zigzag, slash!
I sizzle, SIZZLE, when I dribble,
I'm lightning on the court.
My team calls me The Giant,
Even though I'm kinda short.

The other team might laugh to see
A player tiny as a flea.

But I'm a rocket, fiery hot,
Watch me soar, SOAR, on my jump shot!

Stretching, flexing, push, push, PUSH,
My ball flies up and in—Swoosh Woosh!

I show them all
You don't need tall
To rule the ball!

MM Productions/Corbis

Captain's Log,
May 12, 1868

We set sail from a port in Spain,

Sun high, no sign of rain.

The sea was satin, so blue—so blue.

Our ship was a bird, we flew—we flew.

Just past noon, how very weird,

Came a sound that we most feared.

Thunder rumbled, a giant drum.

Thunder rumbled, rum tum tum.

Rain was pouring, pouring.

The wind was a monster, roaring, roaring.

My crew, extremely terrified,

Froze at their posts, pale and wide-eyed.

A huge wave lifted up our ship,
My feet began to slip, slip, slip.
I knew that it was up to me,
To guide us through that stormy sea.

I grabbed a rope, reached for the mast,
And got back to the helm at last - at last
Shook off the rain, looked at my crew,
"Steady lads, I'll get us through."

The crew heard my call,
Each lad stood up tall.
All hands now on deck, we trimmed every sail.
Courageous, together, we rode out that gale.

Make Connections

Talk about how the person in each poem is inspiring. **ESSENTIAL QUESTION**

In the poems, which person is most inspiring to you? Why? **TEXT TO SELF**

Narrative and Free Verse

Narrative poetry: • Tells a story. • Often has stanzas, or groups of lines. • Often rhymes.

Free verse poetry: • Does not always rhyme. • Can have stanzas with different numbers of lines. • Can tell a story or express a poet's feelings.

🔍 Find Text Evidence

I can tell that "Captain's Log" is a narrative poem. It is a story of a ship's captain who inspires his crew during a bad storm.

page 320

Captain's Log,
May 12, 1868

We set sail from a port in Spain,
Sun high, no sign of rain.
The sea was satin, so blue—so blue.
Our ship was a bird, we flew—we flew.

Just past noon, how very weird,
Came a sound that we most feared.
Thunder rumbled, a giant drum.
Thunder rumbled, rum tum tum.

Rain was pouring, pouring.
The wind was a monster, roaring, roaring.
My crew, extremely terrified,
Froze at their posts, pale and wide-eyed.

320

"Captain's Log" is a narrative poem that rhymes and has stanzas. It tells a story. This part describes the storm and how scared the crew on the ship was.

COLLABORATE

Your Turn

Reread "The Giant." Explain why it is a free verse poem.

Theme

The theme is the main message or lesson in a poem. The details in a poem can help you figure out the theme.

 Find Text Evidence

All the poems in this week are about inspirational people, but each poem has a different theme. I'll reread "The Giant" and look for details. I can use the details to figure out the theme.

Detail
I sizzle when I dribble and I'm lightning on the court.

↓

Detail
The other team might laugh to see a player so small.

↓

Detail

↓

Detail

↓

Theme
If you *believe* in yourself, you can do anything.

 COLLABORATE

Your Turn

Reread "The Giant." Find more details and list them in your graphic organizer. Make sure they support the theme.

Go Digital!
Use the interactive graphic organizer

Repetition and Rhyme

Repetition means that words or phrases in a poem are repeated. A **rhyme** is two or more words that end with the same sounds, such as *pouring* and *roaring*.

Find Text Evidence

Reread "Captain's Log" on pages 320–321. Listen for words or phrases that are repeated. Think about why the poet uses repetition.

page 320

Captain's Log,
May 12, 1868

We set sail from a port in Spain,
Sun high, no sign of rain.
The sea was satin, so blue—so blue.
Our ship was a bird, we flew—we flew.

In the first stanza, the poet repeats the words so blue *and* we flew. *These words also rhyme. This repetition gives the poem a musical quality. It helps me feel the waves and how the ship moves on the sea.*

Your Turn

Reread "Captain's Log." Find examples of repetition and rhyme.

Metaphor

A metaphor compares two things that are very different. It helps you picture, or visualize. "His teeth are white pearls" is a metaphor. It compares teeth to pearls. This metaphor helps me picture bright, white teeth.

Find Text Evidence

On page 318, I read that "Ginger's fingers are shooting stars." This is a metaphor. It compares the way Ginger's fingers move and sign to shooting stars. This metaphor helps me picture Ginger's fingers moving quickly and quietly.

page 318

Ginger's fingers are shooting stars,
They talk of adventurous trips to Mars.
Fingers talking without words,
Signing when sounds can't be heard.

COLLABORATE

Your Turn

Reread the poem "Ginger's Fingers." Find another metaphor. What two things are compared? Talk about how the metaphor helps you visualize.

Readers to ...

Writers choose strong, descriptive words to make their writing interesting and clear. Strong words show, rather than tell. Reread part of "The Giant" below.

Strong Words

Identify **strong words**. How do they help you visualize the way the narrator moves?

Expert Model

Dodge, dart, dash,

Zigzag, slash!

I sizzle, SIZZLE, when I dribble,

I'm lightning on the court.

My team calls me The Giant,

Even though I'm kinda short.

Writers

Anne wrote a poem about someone who inspires her. Read her revisions.

Editing Marks

≡ Make a capital letter.

/ Make a small letter.

⊙ Add a period.

∧ Add.

⤴ Take out.

Grammar Handbook

Irregular Verbs
See page 486.

Student Model

My Grandpa

He walks slower now

and with great care.

                    twinkle
but his eyes still ∧shine

And he likes to share.

            is
My grandpa ∧was the one

I most admire.

   has
He ∧have the biggest Ḧeart

and never seems to tire.

Your Turn

☑ Identity strong words.

☑ Identify an irregular verb.

☑ Tell how revisions improved the poem.

Go Digital!
Write online in Writer's Workspace

327

Unit 5

Take Action

The Big Idea

What are ways people can take action?

Ben Franklin's STOVE

Back in 1742,
Snow fell fast and cold winds blew.

Ben Franklin's fireplace was bright,
Burning wood both day and night.

But Ben would rub his icy feet,
"Too much smoke, and where's the heat?

A fireplace does little good,
For it uses too much wood.

I'll build an iron stove instead,
It will burn less wood," Ben said.

Day after day, Ben's hammer rang
On that iron—CLINK, CLUNK, CLANG.

The stove Ben made helped hot air rise,
With no smoke to sting his eyes.

Now through the winter Ben could boast,
"My stove keeps me warm as toast!"

— **By Charles Ashton**

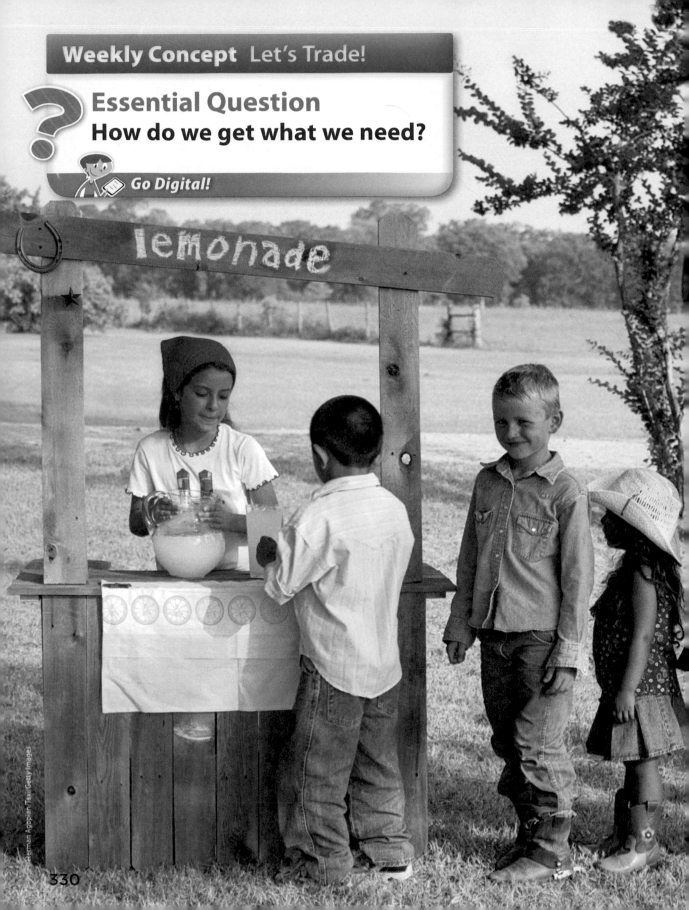

Essential Question
How do we get what we need?

Go Digital!

Give and Take

Everyone is thirsty. All the kids want a glass of Sue's lemonade. They wait in line for what they want and need. They will pay for it in different ways.

▶ Some will use money.

▶ Some will barter, or trade.

▶ Sue is getting what she needs, too. She is working to save money to buy a new bike.

Talk About It

Write words you have learned about getting what you need. Talk with a partner about how to get what you need.

Get What You Need

Vocabulary

Use the picture and the sentence to talk with a partner about each word.

admit

Josh had to **admit** to his mom that he got dirt on her clean sheets.

Tell about something you had to admit.

barter

Amy likes to **barter**, or trade, parts of her lunch with Kim.

What is another word for barter?

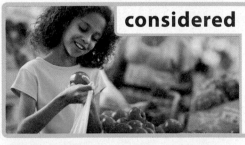

considered

Marta thought carefully as she **considered** which tomato to buy.

Name something you considered doing.

creation

Ella admired her **creation** in art class.

Tell about a creation you made.

humble

My grandfather's house is **humble**, simple, and plain.

What is the opposite of humble?

magnificent

Mr. Jacobs took a picture of the **magnificent** canyon.

What is a magnificent place you know?

payment

Mom gave our neighbor **payment** for the vase.

What do people usually use for payment?

reluctantly

The goats stepped **reluctantly** down the steep path.

Show how you would raise your hand reluctantly.

Your Turn

COLLABORATE

Pick three words. Then write three questions for your partner to answer.

Go Digital! *Use the online visual glossary*

(t) Picture Contact BV/Alamy; (tc) Digital Zoo/Photodisc/Getty Images; (bc) Blend Images/Ariel Skelley/the Agency Collection/Getty Images; (b) Art Wolfe/Stone/Getty Images

Juanita and the Beanstalk

? Essential Question

How do we get what we need?

Read about what Juanita does to get what she needs.

Juanita lived in a small, **humble** cottage with her Mamá and her pet goat, Pepe.

One day Mamá said, "There has been no rain, and our garden has dried up. Juanita, you must go to town and sell your goat. Use the money you get as **payment** to buy some food."

"I don't want to sell Pepe!" cried Juanita. She petted the goat lovingly. But she was an obedient girl and would not disobey her mother. **Reluctantly**, she took Pepe to town. On her way she met an old man who patted Pepe kindly.

"He is for sale," said Juanita with tears in her eyes.

The man replied, "I have no money, but I have some special *frijoles*. If you plant these beans you will never go hungry again. We can **barter**, and I will trade you these beans for your goat."

Chris Vallo

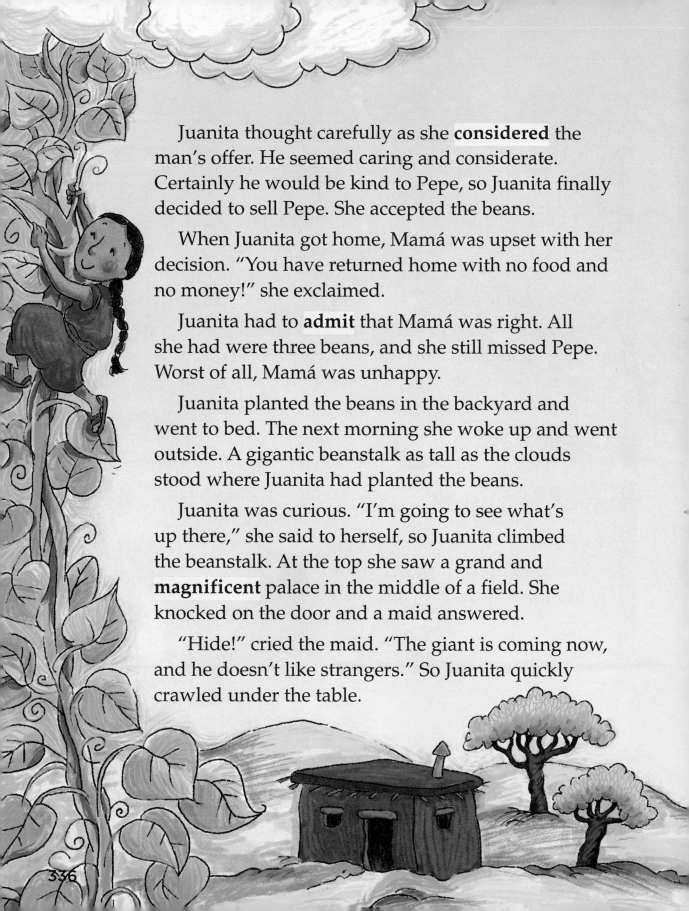

Juanita thought carefully as she **considered** the man's offer. He seemed caring and considerate. Certainly he would be kind to Pepe, so Juanita finally decided to sell Pepe. She accepted the beans.

When Juanita got home, Mamá was upset with her decision. "You have returned home with no food and no money!" she exclaimed.

Juanita had to **admit** that Mamá was right. All she had were three beans, and she still missed Pepe. Worst of all, Mamá was unhappy.

Juanita planted the beans in the backyard and went to bed. The next morning she woke up and went outside. A gigantic beanstalk as tall as the clouds stood where Juanita had planted the beans.

Juanita was curious. "I'm going to see what's up there," she said to herself, so Juanita climbed the beanstalk. At the top she saw a grand and **magnificent** palace in the middle of a field. She knocked on the door and a maid answered.

"Hide!" cried the maid. "The giant is coming now, and he doesn't like strangers." So Juanita quickly crawled under the table.

The giant stomped in carrying an unhappy hen in a cage. He said, "Lay, hen, lay!" Juanita's curiosity grew, and she peeked from under the table. Then she saw the hen's **creation**. Juanita gasped. It was a golden egg!

The poor hen reminded Juanita of Pepe. She wanted to give it a better home. She ran between the giant's legs and grabbed the cage. She raced to the beanstalk. The giant roared in anger and chased after her. Juanita was able to slide down the beanstalk, but the giant was too heavy. He caused the stalk to break and crash to the ground. The beanstalk was gone forever, and Juanita and the hen were safe.

The hen was happy to have a new home and laid many golden eggs. Mamá was happy to use the eggs to buy everything they needed. And Juanita was happy because she was able to trade a golden egg with the old man to get Pepe back!

Make Connections

? How does Juanita get what she needs? **ESSENTIAL QUESTION**

What are some ways you can get what you need? **TEXT TO SELF**

Chris Vallo

337

CCSS **Comprehension Strategy**

Summarize

When you summarize, you retell the most important events in a story. Use details to help you summarize "Juanita and the Beanstalk".

Find Text Evidence

Why does Juanita have to sell her pet goat? Identify important story events. Summarize them in your own words.

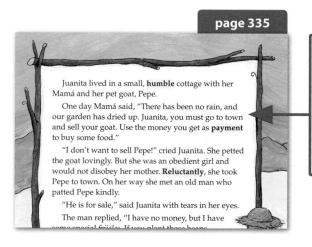

page 335

Juanita lived in a small, **humble** cottage with her Mamá and her pet goat, Pepe.

One day Mamá said, "There has been no rain, and our garden has dried up. Juanita, you must go to town and sell your goat. Use the money you get as **payment** to buy some food."

"I don't want to sell Pepe!" cried Juanita. She petted the goat lovingly. But she was an obedient girl and would not disobey her mother. **Reluctantly**, she took Pepe to town. On her way she met an old man who patted Pepe kindly.

"He is for sale," said Juanita with tears in her eyes.

The man replied, "I have no money, but I have

I read that it hadn't rained and Mamá's garden dried up. They needed money for food. Mamá told Juanita to sell Pepe, her pet goat. These details help me summarize. Juanita had to sell her goat to get money for food.

Your Turn

Reread "Juanita and the Beanstalk." Summarize the most important events that tell how Juanita found the giant's palace.

Chris Vallo

Point of View

A character often has thoughts about other characters or events in a story. This is the point of view. Look for details to figure out the character's point of view.

 ### Find Text Evidence

What does Juanita think about Pepe? I can reread what she does and says. These details will help me figure out Juanita's point of view about the goat.

Details
Juanita tells Mamá that she does not want to sell Pepe.
She pets the goat lovingly.

↓

Point of View

Your Turn COLLABORATE

Reread "Juanita and the Beanstalk." Write details about Juanita's feelings for Pepe in the graphic organizer. Figure out her point of view. Do you agree with Juanita's point of view?

Go Digital!
Use the interactive graphic organizer

Fairy Tale

"Juanita and the Beanstalk" is a fairy tale. A **fairy tale**:

- Is a made-up story with events that could not really happen
- Usually has magical characters or settings
- Almost always has a happy ending with a message

Find Text Evidence

I can tell that "Juanita and the Beanstalk" is a fairy tale. A huge beanstalk could not grow overnight in real life. There are also magical characters, such as a giant and a hen that lays golden eggs. The story has a happy ending, too.

page 337

The giant stomped in carrying an unhappy hen in a cage. He said, "Lay, hen, lay!" Juanita's curiosity grew, and she peeked from under the table. Then she saw the hen's **creation**. Juanita gasped. It was a golden egg!

The poor hen reminded Juanita of Pepe. She wanted to give it a better home. She ran between the giant's legs and grabbed the cage. She raced to the beanstalk. The giant roared in anger and chased after her. Juanita was able to slide down the beanstalk, but the giant was too heavy. He caused the stalk to break and crash to the ground. The beanstalk was gone forever, and Juanita and the hen were safe.

The hen was happy to have a new home and laid many golden eggs. Mamá was happy to use the eggs to buy everything they needed. And Juanita was happy because she was able to trade a golden egg with the old man to get Pepe back!

Make Connections

How does Juanita get what she needs? ESSENTIAL QUESTION

What are some ways you can get what you need? TEXT TO SELF

337

Important **events** in a fairy tale could not really happen.

Fairy tales usually have a happy ending with a message.

Your Turn

With a partner, find two details that show this is a fairy tale. How is it like other fairy tales you know?

COLLABORATE

Root Words

A root word is the simplest form of a word. When you read an unfamiliar word, look for the root word. Then use the root word to figure out what the word means.

 Find Text Evidence

On page 336 of "Juanita and the Beanstalk," I see the word considerate. *I think the root word of* considerate *is* consider. *I know that to* consider *something means "to think about it." Being* considerate *means "thoughtful of others' feelings."*

Juanita thought carefully as she considered the man's offer. The man seemed caring and considerate.

Your Turn

Find the root words. Use them to figure out the meanings of the following words.

decision, *page 336*
curiosity, *page 337*

Chris Vallo

341

Readers to . . .

Writers use different kinds of sentences that make stories more interesting to read. Reread the passage from "Juanita and the Beanstalk."

Vary Sentence Structures

Find different kinds of sentences. How do they make the story more interesting to read?

Expert Model

When Junaita got home, Mamá was upset at her decision. "You have returned home with no food and no money!" she exclaimed.

Juanita had to admit that Mamá was right. All she had were three beans, and she still missed Pepe. Worst of all, Mamá was unhappy.

Chris Vallo

Writers

Stephanie wrote a story about a trade she made. Read her revisions.

Student Model

A Perfect Trade

My friend Jimmy and I love animals⊙ Jimmy and me We have many books about animals. Jimmy has an interesting book about giraffes. I have a great book about sharks. One day, I trading traded my shark book for his giraffe book. He and I We both got to read new books. How do we feel about trading? We think it is great!

Your Turn

COLLABORATE

☑ Identify different kinds of sentences.
☑ Identify singular and plural pronouns.
☑ Tell how revisions improved the writing.

Go Digital!
Write online in Writer's Workspace

Essential Question
How can we reuse what we already have?

Go Digital!

Bob Elsdale/Stone/Getty Images

Recycling

Old tires can be reused. They make great swings. They also can be used to build strong walls on land and to make reefs for animals under water.

▶ Almost everything we use can be recycled.

▶ Reusing things helps keep our Earth clean.

▶ We all should reuse and recycle what we can.

Talk About It

Write words you have learned about recycling. Talk with a partner about ways you can reuse and recycle.

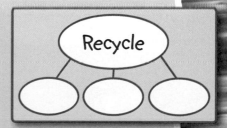

Recycle

Vocabulary

Use the picture and the sentence to talk with a partner about each word.

conservation

Kara's light bulb saves energy and is a good example of **conservation**.

What is another example of conservation?

discouraged

Shawn felt **discouraged** when he and his mom couldn't find his lost football.

What would make you feel discouraged?

frustration

Diane gasped out of **frustration** when her computer screen went blank.

Tell about a time you showed frustration.

gazed

Tory **gazed** at the stars through her telescope.

When have you gazed at the stars?

jubilant

Ryan's team felt **jubilant** and delighted when they won the contest.

What word means almost the same as jubilant?

recycling

Sheila is **recycling** empty containers.

What are other things you could be recycling?

remaining

These seven puppies are **remaining** after we gave two away.

How many days of school are remaining this week?

BABY BULLDOGS

tinkered

Jon and his dad **tinkered** with the bike and got it working again.

Think of another word for tinkered.

Your Turn

COLLABORATE

Pick three words. Write three questions for your partner to answer.

Go Digital! *Use the online visual glossary*

The New HOOP

Chris Vallo

? Essential Question

How can we reuse what we already have?

Read to see how Kim and Marco reuse something to solve their problem.

348

Marco **gazed** at the basketball hoop and threw the ball up. It whizzed through the air. "Score!" he shouted as the ball fell through with a swish.

"You won this time, but I'll beat you next time, Marco!" said Kim as the two friends made their way home. "I wish we could play at home, too, instead of only at school. It's not fair." The basketball hoop in their neighborhood park had been ruined when a tree fell and crushed it.

"My dad says the Parks Department doesn't have enough money to buy a new hoop yet," grumbled Marco in **frustration**.

"I feel so **discouraged**," said Kim. "I guess there's nothing we can do."

Marco and Kim walked past the city's **recycling** center. They waved at the manager, Mr. Morse. His job was to separate the plastic, paper, and metal items people brought to him. He was transferring cardboard from an overflowing bin into large, empty containers.

Marco stared at all the old stuff. "That gives me an idea!" he said. "Mr. Morse, do you have anything we could reuse to make a basketball hoop?"

Mr. Morse picked up a plastic laundry basket. "We were going to recycle this basket, but I think it's reusable."

"It looks useless, old, and cracked," said Kim.

"No, it could be useful," said Marco. "We can cut off the bottom to make a fine hoop, and then an adult can help us attach it to a post."

Kim frowned. "I want a new basketball hoop," she said. "Not someone else's hand-me-down."

"Why?" wondered Marco. "Reusing things is a great way to practice **conservation**. It stops waste."

"I guess we can try," said Kim. "But I still don't believe it will be as good as a new one."

They took the basket to Marco's house. His older brother, Victor, got some leftover wood from an old building project. Together they **tinkered** with the materials and made a post and a backboard.

When Marco went to attach the basket to the backboard, he found his two cats napping in it. "I see someone has found a way to reuse the basket already!" he laughed. He let them sleep a few minutes longer.

Chris Vallo

When all the parts were ready, there was only one thing **remaining** to do. Marco, Kim, and Victor took everything to the park. Kim helped dig the hole for the post, but she was still unsure. Next, Marco helped Victor ease the backboard and basket carefully into the hole.

"It looks better than I thought it would!" said Kim.

"Here's the real test!" grinned Marco. He tossed her the basketball. Kim bounced the ball, aimed, and shot a perfect basket. She was **jubilant**.

"Wow, I was wrong," she said. "This recycled basketball hoop is really great. Now we can play whenever we want!"

"Yes, and I can beat you whenever I want," grinned Marco.

"Oh, no you can't!" laughed Kim. The two friends played basketball until dinner time.

Make Connections

What problem do Kim and Marco have? How do they reuse something to solve it? **ESSENTIAL QUESTION**

Discuss how you reused something to solve a problem. How did it work? **TEXT TO SELF**

Summarize

When you summarize, you retell the most important events in the story. Use events to help you summarize "The New Hoop."

 Find Text Evidence

You may not know why Kim and Marco feel frustrated and discouraged. Reread and identify important story events. Then summarize them in your own words.

---

page 349

Marco **gazed** at the basketball hoop and threw the ball up. It whizzed through the air. "Score!" he shouted as the ball fell through with a swish.

"You won this time, but I'll beat you next time, Marco!" said Kim as the two friends made their way home. "I wish we could play at home, too, instead of only at school. It's not fair." The basketball hoop in their neighborhood park had been ruined when a tree fell and crushed it.

"My dad says the Parks Department doesn't have enough money to buy a new hoop yet," grumbled Marco in **frustration**.

"I feel so **discouraged**," said Kim. "I guess there's nothing we can do."

I read that <u>Kim said she wished she could play ball at home. The basketball hoop was broken, and there was no money to fix it. Marco grumbled. Kim said she felt discouraged.</u> These details help me summarize. Kim and Marco are frustrated because they can't play basketball in their neighborhood.

Your Turn

COLLABORATE

Reread "The New Hoop." Summarize the most important events that tell how Kim and Marco solve their problem.

Point of View

Point of view is what a character thinks about other characters or events in a story. Look at Marco's actions and words to figure out his point of view.

Find Text Evidence

What does Marco think about reusing recycled things? I can reread what he does and says. These details will help me figure out his point of view.

Details
Marco asks Mr. Morse if he has anything they could reuse to make a basketball hoop.

↓

Point of View

Your Turn

COLLABORATE

Reread "The New Hoop." Write about Marco's feelings about reusing things in your graphic organizer. Figure out his point of view. Do you agree with Marco's point of view about recycling? Why or why not?

Go Digital!
Use the interactive graphic organizer

Realistic Fiction

"The New Hoop" is realistic fiction. **Realistic fiction**:

- Is a made-up story that could really happen
- Has illustrations that give information about the characters, setting, and events

Find Text Evidence

I can tell that "The New Hoop" is realistic fiction. Marco and Kim have a problem to solve. That could happen in real life. There are also illustrations that give details about the characters and setting.

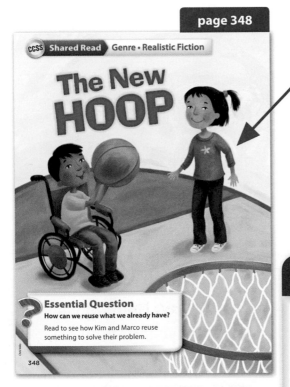

page 348

CCSS Shared Read ▸ Genre • Realistic Fiction

The New HOOP

Essential Question
How can we reuse what we already have?
Read to see how Kim and Marco reuse something to solve their problem.

348

Illustrations Illustrations give you more details about the characters, setting, or events in a story.

Your Turn

Find two things in the story that could happen in real life. Talk about why "The New Hoop" is realistic fiction.

Homographs

Homographs are words that are spelled the same but have different meanings. They are sometimes pronounced differently. Use nearby words as clues to help you figure out the meaning of a homograph.

 Find Text Evidence

On page 349 in "The New Hoop" I see the word can. *I know* can *means "to be able to" or "a metal container." I will reread sentences from the story to figure out which meaning fits. In this sentence,* can *means "to be able to."*

"I feel so discouraged," said Kim. " I guess there's nothing we can do."

Your Turn

COLLABORATE

Use context clues to figure out the meaning for the following homographs.

fair, *page 349*
fine, *page 350*
post, *page 350*

Chris Vallo

Readers to...

Writers choose words to tell how something looks, sounds, smells, tastes, or feels. Sensory language helps describe events in a clear way. Reread from "The New Hoop."

Sensory Language

Find examples of **sensory language**. How do they help describe an event?

Expert Model

Marco gazed at the basketball hoop and threw the ball up. It whizzed through the air. "Score!" he shouted as the ball fell through with a swish.

"You won this time, but I'll beat you next time Marco!" said Kim as the two friends made their way home. "I wish we could play at home, too, instead of only at school. It's not fair." The basketball hoop in their neighborhood park had been ruined when a tree fell and crushed it.

Chris Vallo

Writers

Jacob wrote a story about something he reused. Read Jacob's revisions.

Editing Marks

≡ Make a capital letter.

／ Make a small letter.

⊙ Add a period.

∧ Add.

⌵ Take out.

Grammar Handbook

Subject and Object Pronouns
See page 488.

Student Model

Mom's Birthday Gift

It was Mom's birthday. We

had lots of ^clean and sturdy^ cardboard boxes in the

recycling bin. ~~Me~~ ^I^ wanted to make

~~my mother~~ ^her^ a gift from a box. First

I cut a frame from the front of the

box. ^Then^ I glued ^bright red and white^ buttons and ribbons

on it. After it dried, I placed the

photo inside. My mom opened the

gift and gasped loudly. ~~s~~he gave me

^great, big^ a hug. I guess she liked my gift⊙

Your Turn

☑ Identify sensory language.
☑ Identify subject and object pronouns.
☑ Tell how revisions improved the writing.

Go Digital!
Write online in Writer's Workspace

357

Essential Question
How do teams work together?

 Go Digital!

Peter Kneffel/dpa/Corbis

358

TEAMWORK

Sasha is trained to rescue people. She and her team work together to search for and help people lost in the snow.

► There are many different kinds of teams.

► Team members must trust each other and communicate well.

► Teamwork gets the job done.

Talk About It

Write words you have learned about teamwork. Talk with a partner about how teams work together.

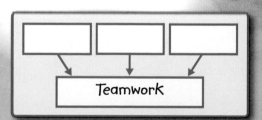

Teamwork

Vocabulary

Use the picture and the sentence to talk with a partner about each word.

accidental

Jason felt bad about the **accidental** mess he made on the driveway.

What does the word accidental mean?

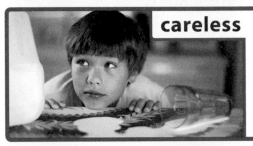

careless

Harry was **careless** and not paying attention when he spilled his milk.

What is the opposite of careless?

disasters

Tornadoes and other natural **disasters** often cause a lot of damage.

What are some other natural disasters you have heard about?

equipment

Firefighters wear special **equipment** to fight fires.

What kind of equipment do firefighters need to do their jobs?

harmful

Poison ivy leaves may be **harmful** to your skin if you are allergic to them.

What other things might be harmful to you?

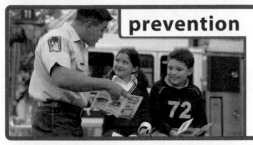

prevention

Dave is teaching Fran and Juan about fire **prevention**.

Why should you learn about fire prevention?

purpose

The **purpose** of Jay's helmet is to protect his head.

What purpose do gloves have?

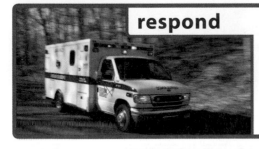

respond

It is important for this ambulance to **respond** quickly to an emergency.

Why is it important to respond quickly to an emergency?

Your Turn

COLLABORATE

Pick three words. Write three questions for your partner to answer.

Go Digital! *Use the online visual glossary*

Rescue Dogs Save the Day

Essential Question

How do teams work together?

Read how rescue dogs help in emergencies.

Rescue dogs are trained to go anywhere they are needed.

Rescue teams are there when we need them. They **respond** quickly to help people in trouble. They are brave heroes. But heroes aren't always people. Heroes can be dogs, too!

Rescue Dogs Are Heroes

Rescue dogs are always ready to go to work. They team up with police, fire and other rescue workers. They are good at finding people who are lost. They rescue families after earthquakes and other **disasters**. They work in all types of weather. And the best news is that rescue dogs can do their jobs with no special **equipment**. All they need is their excellent hearing and a good nose!

Rescue dogs are smart and brave. They listen well to commands and do their jobs even when they are tired, thirsty, or hungry. They are friendly and get along well with their handlers, the people who work with them. They also must be obedient and do what they are told.

Certain breeds of dogs are easier to train to work in dangerous rescue situations than others. The Border Collie is one breed of dog used during disasters and emergencies. Border collies can work for a long time. They do not get tired easily, and that's important. But dogs need more than energy. They also need to follow commands, and that takes a lot of training.

Getting Ready to Work

Rescue dogs begin their training as puppies. It can take up to two years to completely train a rescue dog. Then it is able to save people in **harmful** and dangerous situations.

The dogs learn to work outdoors in heat, cold, and bad weather. They run, jump, and climb for many hours every day. Rescue dogs also learn to ignore everything around them while they are working. This helps them to focus on the job and keeps them from making **careless** mistakes.

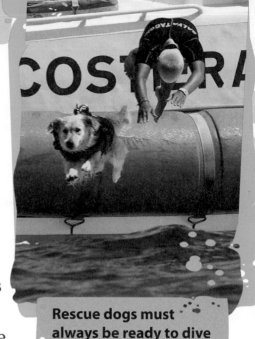

Rescue dogs must always be ready to dive in and help someone.

Everything a rescue dog learns to do has a **purpose**. Even friendship is not **accidental**. A dog and the people it works with must learn to communicate as a team. They trust each other. And when they have practiced and trained enough, they are ready to participate, or take part in, a real rescue mission.

Best Rescue Dog Breeds

These dogs make great rescue dogs.				
Dog Breed	**Labrador Retriever**	**German Shepherd**	**Bloodhound**	**Border Collie**
Rescue Trait	friendly	brave	great sense of smell	lots of energy and stamina

364

Good Dog!

When a hiker is lost, a rescue dog sniffs the air and the ground to find her. A dog's sense of smell is much stronger than a person's. Rescue dogs can even smell someone trapped under fifteen feet of snow. When a dog finds someone, it barks to alert its partner. The rescue worker trusts the dog, so the team works quickly to save a life. At the end of every rescue, the dog gets praise and treats for doing a great job.

Sometimes rescue teams go to schools to teach children about safety and disaster **prevention**. They show children how to stay safe and what to do during emergencies. This job is fun for rescue dogs. They get lots of attention for just doing what they do best – helping people. Rescue dogs really are heroes!

This team works on snowy mountains.

Make Connections

How do rescue workers and dogs work together in an emergency? **ESSENTIAL QUESTION**

What do you think would be the best thing about working with a rescue dog? **TEXT TO SELF**

Tom Bear/Aurora/Getty Images

Ask and Answer Questions

Stop and ask yourself questions as you read. Then reread to find details to support your answers.

 Find Text Evidence

Reread the section "Rescue Dogs Are Heroes" on page 363. Think of a question and then read to answer it.

> **page 363**
>
> Rescue teams are there when we need them. They **respond** quickly to help people in trouble. They are brave heroes. But heroes aren't always people. Heroes can be dogs, too!
>
> ## Rescue Dogs Are Heroes
>
> Rescue dogs are always ready to go to work. They team up with police, fire and other rescue workers. They are good at finding people who are lost. They rescue families after earthquakes and other **disasters**. They work in all types of weather. And the best news is that rescue dogs can do their jobs with no special **equipment**. All they need is their excellent hearing and a good nose!
>
> Rescue dogs are smart and brave. They listen well to commands and do their jobs even when

I have a question. What do rescue dogs do? I read that rescue dogs find people who are lost and rescue families after disasters. They work in all kinds of weather. Now I can answer my question. Rescue dogs work hard to save lives.

Your Turn

COLLABORATE

Reread "Getting Ready to Work." Think of a question. You might ask: How do dogs train to be rescue dogs? Reread the section to find the answer.

Frank Leonhardt/dpa/Corbis

Author's Point of View

A point of view is what an author thinks about a topic. Look for details that show what the author thinks. Decide if you agree with the author.

 Find Text Evidence

What does the author think about rescue dogs? I can reread and look for details that tell me what the author thinks. This will help me figure out the author's point of view.

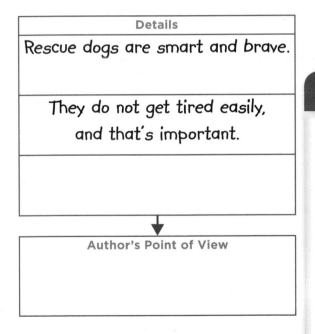

Details
Rescue dogs are smart and brave.
They do not get tired easily, and that's important.

↓

Author's Point of View

Your Turn COLLABORATE

Reread "Rescue Dogs Save the Day." Find details that tell what the author thinks about rescue dogs. What is the author's point of view? List details and point of view in your graphic organizer. Do you agree with the author's point of view? Why or why not?

Go Digital!
Use the interactive graphic organizer

Expository Text

"Rescue Dogs Save the Day" is an expository text.

Expository text:
- May explain a science topic
- Includes text features such as headings, photographs, captions, and a chart

Find Text Evidence

I can tell that "Rescue Dogs Save the Day" is an expository text. It gives facts and information about rescue dogs. It has photographs with captions. It also has headings and a chart.

page 364

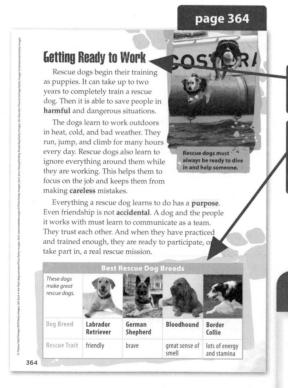

Getting Ready to Work

Rescue dogs begin their training as puppies. It can take up to two years to completely train a rescue dog. Then it is able to save people in **harmful** and dangerous situations.

The dogs learn to work outdoors in heat, cold, and bad weather. They run, jump, and climb for many hours every day. Rescue dogs also learn to ignore everything around them while they are working. This helps them to focus on the job and keeps them from making **careless** mistakes.

Everything a rescue dog learns to do has a **purpose**. Even friendship is not **accidental**. A dog and the people it works with must learn to communicate as a team. They trust each other. And when they have practiced and trained enough, they are ready to participate, or take part in, a real rescue mission.

Rescue dogs must always be ready to dive in and help someone.

Best Rescue Dog Breeds			
These dogs make great rescue dogs.			
Dog Breed	Labrador Retriever	German Shepherd	Bloodhound	Border Collie
Rescue Trait	friendly	brave	great sense of smell	lots of energy and stamina

364

Text Features

Headings A heading tells what a section of text is about.

Chart A chart presents information about the topic in a way that is easy to read.

COLLABORATE

Your Turn

Look at the chart on page 364. Why do German Shepherds make good rescue dogs?

Sentence Clues

As you read, you may come across a word you don't know. Look at other words in the same sentence. They can give you clues about the word's meaning.

 Find Text Evidence

On page 363, I see the word handlers. *I'm not sure what that word means. I see the words "people who work with them" in the same sentence. This helps me figure out that* handlers *are "the people who work with rescue dogs."*

They are friendly and get along well with their handlers, the people who work with them.

Your Turn

Find sentence clues. Use them to help you figure out the meanings of these words.

obedient, *page 363*

participate, *page 364*

Readers to...

Writers group related ideas together. A strong paragraph has a topic sentence that states the main idea and supporting sentences that give details. Reread the passage from "Rescue Dogs Save the Day."

Expert Model

Strong Paragraph

Find the topic sentence. What details support the topic sentence?

Certain breeds of dogs are easier to train to work in dangerous rescue situations than others. The Border Collie is one breed of dog used during disasters and emergencies. Border collies can work for a long time. They do not get tired easily, and that's important. But dogs need more than energy. They also need to follow commands, and that takes a lot of training.

Writers

Katie wrote about how emergency workers help people. Read Katie's revision.

Editing Marks

≡ Make a capital letter.
/ Make a small letter.
⊙ Add a period.
∧ Add
ℛ Take out.

Grammar Handbook

Pronoun-Verb Agreement
See page 489.

Student Model

Firefighters Help Us

Firefighters help our community. They help put out fires. They ~~does~~ **do** dangerous work. ∧ ~~I saw a firefighter once.~~ When you call firefighters, they come very quickly. They work fast to put out a fire⊙ I ~~thinks~~ **think** firefighters are brave and helpful.

Your Turn

COLLABORATE

☑ Identify the topic sentence.
☑ Identify a pronoun-verb agreement.
☑ Tell how revisions improved the writing.

Go Digital!
Write online in Writer's Workspace

David L. Moore–Oahu/Alamy

372

CITIZENSHIP

Lou is helping his town get ready to honor its heroes. He is being a good citizen by participating in something that is important to his community.

- ▶ Being a good citizen means helping other people.
- ▶ It means following the rules and being respectful of others.
- ▶ Good citizenship helps make a community safe.

Talk About It

Write words you have learned about citizenship. Talk about ways you can be a good citizen.

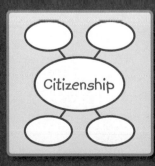

Vocabulary

Use the picture and the sentence to talk with a partner about each word.

citizenship

Planting a tree in your community is an example of good **citizenship**.

What is another example of good citizenship?

continued

Justin **continued** to read his book all afternoon.

What is the opposite of continued?

daring

One brave penguin made a **daring** dive into the cold sea.

Tell about something daring you have seen.

horrified

Paul and his mother were **horrified** by the scary movie.

What does it mean to feel horrified?

participate

Barb and her friends like to **participate** in sack races at the picnic.

What are some games you like to participate in?

proposed

Mom **proposed** that they look online to find the answer to Tina's question.

Tell about something you proposed to your family.

unfairness

The baseball player discussed the **unfairness** of the referee's call.

What word means the opposite of unfairness?

waver

Ted's confidence started to **waver** when he forgot the answer.

Show how you would look if you start to waver.

Your Turn

COLLABORATE

Pick three words. Write three questions for your partner to answer.

Go Digital! *Use the online visual glossary*

Dolores Huerta
GROWING UP STRONG

? Essential Question

What do good citizens do?

Read how Dolores Huerta's actions helped many people.

Dolores Huerta learned to help people by watching her mother. Good **citizenship** was important to her, and she taught Dolores that women can be strong leaders. When Dolores grew up, she had the same beliefs.

Good Citizens

Dolores was born on April 10, 1930. She lived in a small town in New Mexico until she was three years old. Then she moved to California with her mother and two brothers. Dolores grew up watching her mother **participate** in community organizations. Her mother believed that all people deserved to be treated fairly.

When Dolores was a young girl, her mother owned a hotel and a restaurant. Many farm workers who lived in their town were poor and hungry. They were paid very little for their hard work. Dolores' mother let them stay at her hotel and eat at her restaurant for free. This taught Dolores and her brothers that good citizens get involved in the community by helping their neighbors.

Dolores Huerta helped farm workers who spent many hours working in fields.

Dolores Goes to School

Dolores saw how hard life was for farm workers in California. She wanted everyone to be treated fairly. This attitude **continued** as she attended college and studied to become a teacher.

Many of the students that Dolores taught were the children of farm workers. These students were often tired and hungry. They came to school barefoot because they had no shoes. Dolores knew she needed to help them. As a result, she went to her school's principal and **proposed** some good ideas. She tried to get free lunches and milk for the children. She tried to get them new clothes and shoes.

Trying to help the children was a **daring** thing for Dolores to do. The other teachers did not agree with her ideas. Dolores risked a lot, but her beliefs did not **waver**. She decided to do something about the **unfairness** she saw. She wanted to find a better way to help farm workers and their families.

Dolores: Strong and Fair

This time line shows important dates in Dolores Huerta's life.

1930   1940   1950   1960   1970   1980

1933: Moved to California

1953–1955: Worked as a teacher

1955: Met César Chávez

1930: Dolores is born

1962: Dolores and César started National Farm Worker's Association

1975: Helped pass laws to protect farm workers

Dolores Stands Strong

Every day Dolores saw people working in unusually unsafe and disagreeable conditions. She was **horrified**. Many farm workers had little money to feed their families. Dolores decided to do something.

In 1955, Dolores met César Chávez. He wanted to make life better for farm workers, too. Dolores and César organized the

Dolores Huerta speaks out for farm workers at a rally in 1969.

workers into a group called the National Farm Workers Association. This group protected the rights of the farm workers. It helped make big farms treat them better. As a result, working conditions on the farms improved.

Growing up with a mother who cared about other people taught Dolores to be a good citizen. Her kind and brave acts helped farm workers and their families. Who is a good citizen? Dolores Huerta is!

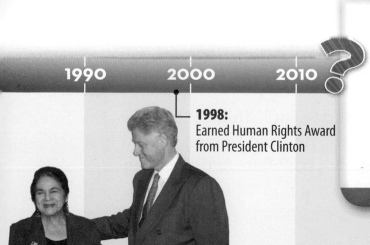

1990   2000   2010

1998:
Earned Human Rights Award from President Clinton

Make Connections

How did Dolores Huerta's actions make her a good citizen? **ESSENTIAL QUESTION**

What can you do to improve people's lives? **TEXT TO SELF**

Ask and Answer Questions

Ask yourself questions as you read. Then read on or reread to find the answers.

 Find Text Evidence

Reread "Good Citizens" on page 377. Think of a question and then read to find the answer.

> **page 377**
>
> **D**olores Huerta learned to help people by watching her mother. Good **citizenship** was important to her, and she taught Dolores that women can be strong leaders. When Dolores grew up, she had the same beliefs.
>
> **Good Citizens**
>
> Dolores was born on April 10, 1930. She lived in a small town in New Mexico until she was three years old. Then she moved to California with her mother and two brothers. Dolores grew up watching her mother **participate** in community organizations. Her mother believed that all people deserved to be treated fairly.
>
> When Dolores was a young girl, her mother owned

I have a question. How did Dolores learn to be a good citizen? I read that <u>when Dolores was young, her mother let the farm workers eat and live for free at her hotel.</u> Now I can answer my question. Dolores learned to be a good citizen by watching her mother.

Your Turn

Reread "Dolores Goes to School." Think of a question. You might ask: How did Dolores try to help the children in her class? Reread the section to find the answer.

Author's Point of View

Point of view is what an author thinks about a topic. Look for details that show what the author thinks. Decide if you agree with the author's point of view.

 Find Text Evidence

What does the author think about Dolores Huerta? I can reread and look for details that tell me what the author thinks. This will help me figure out the author's point of view.

Details
Dolores' mother taught her to be a good citizen and a strong leader.
Helping her students was a daring thing for Dolores to do.

↓

Author's Point of View

Your Turn COLLABORATE

Reread "Dolores Huerta, Growing Up Strong." Find more details that tell how the author feels about Dolores. What is the author's point of view? List them in your graphic organizer. Do you agree with the author's point of view about Dolores Huerta?

Go Digital!
Use the interactive graphic organizer

Biography

"Dolores Huerta, Growing Up Strong" is a biography.
A **biography**:

- Tells the true story of a real person's life
- Is written by another person
- Includes text features such as timelines, photographs, and captions

 Find Text Evidence

I can tell that "Dolores Huerta, Growing Up Strong" is a biography. The author gives facts and information about Dolores Huerta's life. There is also a timeline that shows important events in Dolores' life in time order.

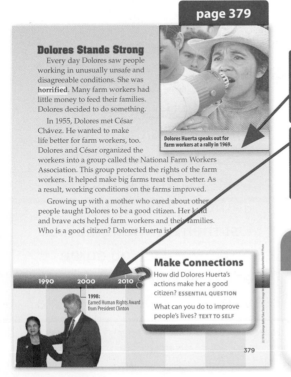

page 379

Dolores Stands Strong

Every day Dolores saw people working in unusually unsafe and disagreeable conditions. She was **horrified**. Many farm workers had little money to feed their families. Dolores decided to do something.

In 1955, Dolores met César Chávez. He wanted to make life better for farm workers, too. Dolores and César organized the workers into a group called the National Farm Workers Association. This group protected the rights of the farm workers. It helped make big farms treat them better. As a result, working conditions on the farms improved.

Growing up with a mother who cared about other people taught Dolores to be a good citizen. Her kind and brave acts helped farm workers and their families. Who is a good citizen? Dolores Huerta is!

Dolores Huerta speaks out for farm workers at a rally in 1969.

1990 2000 2010

1998:
Earned Human Rights Award from President Clinton

Make Connections

How did Dolores Huerta's actions make her a good citizen? ESSENTIAL QUESTION

What can you do to improve people's lives? TEXT TO SELF

379

Text Features

Caption A caption tells about a photograph. It gives information that is not included in the text.

Timeline A timeline shows the time order in which important dates and events happened.

 COLLABORATE

Your Turn

Look at the timeline on pages 378 and 379. In what year did Dolores Huerta meet César Chávez?

Prefixes and Suffixes

A prefix is a word part added to the beginning of a word. A suffix is added at the end. To figure out the meaning of a word with a prefix and suffix, find the root word first.

 Find Text Evidence

On page 379, I see the word unusually. *I find the root word* usual *first. I know the prefix,* un- *means "not," and the suffix* -ly *means "in a way that." The word* unusually *must mean "not in a usual way."*

Everyday Dolores saw people working in unusually unsafe and disagreeable conditions.

COLLABORATE

Your Turn

Find the root word. Then use the prefix and suffix to figure out the meanings of each word.

unfairness, *page 378*

disagreeable, *page 379*

Readers to ...

Writers use questions or fascinating facts to grab the reader's attention. A strong opening states the topic and makes the reader want to keep reading. Reread this passage from "Dolores Huerta, Growing Up Strong."

Expert Model

Strong Opening

Read the **opening**. Why do the first few lines make you want to read more?

Dolores Huerta learned to help people by watching her mother. Good citizenship was important to her and she taught Dolores that women can be strong leaders. When Dolores grew up, she had the same beliefs.

384

Writers

Raj wrote how he feels about being a good citizen. Read Raj's revision.

Editing Marks

≡ Make a capital letter.

/ Make a small letter.

⊙ Add a period.

∧ Add

⤴ Take out.

Grammar Handbook

Possessive Pronouns
See page 489.

Student Model

Be a Good Citizen

It is very important to be a good citizen. ~~Me~~ ^{My} friend Sari is a great citizen. She helps her neighbors, and treat^s everyone fairly⊙ ~~?~~ Sari always follows rules⤴ ^{and} ~~She always~~ participates in school and community activities. It ^{is} ~~was~~ important to be a good citizen like Sari.

Your Turn

COLLABORATE

☑ Identify the strong opening.

☑ Identify possessive pronouns.

☑ Tell how revisions improved the writing.

Go Digital!
Write online in Writer's Workspace

Essential Question
What are different kinds of energy?

Go Digital!

Simon Jarratt/Corbis

We've Got the Power!

Carlos lives near a windmill farm. He loves to ride past on his bike and watch the turbines spin in the wind. Wind power makes electricity that heats his house and powers his lights.

- ▶ Energy comes from many different sources.
- ▶ Energy from the wind and the Sun are renewable. They can be used again and again.

Talk About It

Write words you have learned about energy. Talk with a partner about the different kinds of energy.

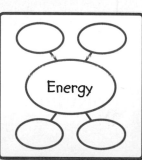

Energy

Vocabulary

Use the picture and the sentence to talk with a partner about each word.

energy

Good food gives Ron and his grandpa lots of **energy** to work together.

Where do cars get their energy?

natural

Cotton is a **natural** material used to make clothes.

Name a natural material that is used in buildings.

pollution

Water **pollution**, such as garbage and chemicals, can harm animals.

Name something that causes air pollution.

produce

The Sun can **produce** enough solar power to heat this family's home.

What word means the same as produce?

renewable

Trees are a **renewable** resource because more will always grow.

What does the word renewable mean?

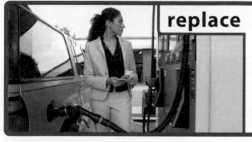

replace

Soon Tina will **replace** her car with one that runs on electricity.

Name something that you can replace.

sources

Wind and solar power are two energy **sources** we can use.

What are your energy sources?

traditional

Cho and her mother like to eat **traditional** foods, such as rice and fish.

What is another word for traditional?

Your Turn

COLLABORATE

Pick three words. Write three questions for your partner to answer.

Go Digital! **Use the online visual glossary**

Essential Question

What are different kinds of energy?

Read why solar energy is a good source of power.

This energetic snowboarder gets his power from the healthful foods he eats.

390

Here Comes Solar Power

What do you have in common with a car and a factory? You both need **energy** to run. Energy keeps things moving.

Energy Today

You get your energy from eating healthful foods. Most factories, homes, and cars get their energy from fossil fuels. Coal, petroleum, and natural gas are fossil fuels. They have been the **traditional**, or usual, energy **sources** for more than a century. Today, most of the energy we use in the United States comes from burning fossil fuels.

But these fuels come from deep under the Earth's surface, and they are running out. They cannot be reused. Once a fossil fuel is gone, it's gone forever. So we need alternative energy sources to **replace** them. Scientists know that there is no other way to keep our country going and growing. So they are looking for new, alternative sources of energy that won't run out.

Cheaper and Cleaner

Solar power is one source of **renewable** energy. And it is not expensive. As a result, many people are placing solar panels on the roofs of homes and large buildings. Solar panels look like giant mirrors, and they capture energy from the Sun.

On a bright day, the Sun's rays hit the solar panel and cause it to **produce** electricity. Then the electricity flows into the building. As a result, there is enough energy to raise the temperature inside homes, and turn on lights, stoves, and computers.

The Future

More companies are turning to solar power to replace fossil fuels. It's **natural**. That means it isn't made, or changed, by people. Solar power is cheaper than fossil fuels, and it does not create **pollution**.

Today there are millions of people around the world using solar power to produce electricity for their homes and businesses. Someday solar power may completely replace fossil fuels.

Solar panels are placed on the roofs of buildings.

GO SUNSHiNe!

Renewable energy is where it's at. And solar power is at the top of our list. Here are the top reasons why solar energy is so hot!

- Solar power is cheaper than fossil fuels.
- It is renewable.
- It doesn't cause pollution and is good for our environment.
- Power from the Sun is always available.
- Solar power is natural.

Solar energy can do just about everything that fossil fuels do. Everyone should use solar power. It's good news for the planet!

Thanks to solar power, Paul can power up and listen to his MP3 player.

Make Connections

Why is solar power a good source of energy? **ESSENTIAL QUESTION**

What are some ways you might use solar power? **TEXT TO SELF**

(r) Cultura Creative/Alamy

Ask and Answer Questions

Ask yourself questions as you read "Here Comes Solar Power." Then look for details to support your answers.

 Find Text Evidence

Look at the section "Energy Today" on page 391. Think of a question. Then read to answer it.

> page 391
>
> ### Energy Today
>
> You get your energy from eating healthful foods. Most factories, homes, and cars get their energy from fossil fuels. Coal, petroleum, and natural gas are fossil fuels. They have been the **traditional**, or usual, energy **sources** for more than a century. Today, most of the energy we use in the United States comes from burning fossil fuels.
>
> But these fuels come from deep under

I have a question. What are fossil fuels? I read that most factories, homes, and cars run on fossil fuels. They come from deep under the Earth's surface and are running out. Now I can answer my question. Fossil fuels come from the Earth and will not always be there.

Your Turn

COLLABORATE

Think of one question about solar power. You might ask: How do solar panels work? Reread page 392 to answer it.

Cause and Effect

A cause is why something happens. An effect is what happens. They happen in time order. Signal words, such as *so, as a result,* and *because* help you find causes and effects.

 Find Text Evidence

On page 391 I read that scientists are looking for alternative sources of energy. This is the effect. Now I can find the cause. Fossil fuels are running out and we need new energy sources.

Cause	→	Effect
First Fossil fuels are running out and we need new energy sources.	→	Scientists are looking for alternative sources of energy.
Next Solar power is renewable and is not expensive.	→	
Then The sun's rays hit the solar panel.	→	

Your Turn

Reread "Here Comes Solar Power." Use signal words to find more causes and effects. Write them in the graphic organizer.

Go Digital!
Use the interactive graphic organizer

Expository Text

"Here Comes Solar Power" is an expository text.

Expository text:
- Often includes facts and information about a science topic.
- Includes text features such photographs, captions, headings, and a sidebar.

 Find Text Evidence

I can tell that "Here Comes Solar Power" is an expository text. It has science facts and information. It has a sidebar that presents how the author feels about solar energy.

page 393

TIME

GO SUNSHINE!

Renewable energy is where it's at. And solar power is at the top of our list. Here are the top reasons why solar energy is so hot!

- Solar power is cheaper than fossil fuels.
- It is renewable.
- It doesn't cause pollution and is good for our environment.
- Power from the Sun is always available.
- Solar power is natural.

Solar energy can do just about everything that fossil fuels do. Everyone should use solar power. It's good news for the planet!

Thanks to solar power, Paul can power up and listen to his MP3 player.

Make Connections
Why is solar power a good source of energy? ESSENTIAL QUESTION

What are some ways you might use solar power? TEXT TO SELF

393

Text Features
Photographs and captions
Photographs and captions give extra information about a topic.

Sidebar A sidebar may present the author's opinion about a topic.

COLLABORATE

Your Turn

Reread the sidebar on page 393. Tell a partner three reasons why people should use solar power.

Homophones

Homophones are words that sound the same but have different meanings and different spellings. The words *sea* and *see* are homophones. Use context clues to figure out a homophone's meaning.

 ## Find Text Evidence

I see the word need *on page 391. Need and* knead *are homophones. Need means "to require something." Knead means "to mix with your hands." The words sound the same, but have different meanings and spellings. I can use context clues to figure out what* need *means. Here it means "to require."*

You both need energy to run.

Your Turn

COLLABORATE

Use context clues to figure out what this word means. Then find its homophone.

rays, *page 392*

Corey Rich/Aurora Open/Corbis

397

Readers to...

Writers use voice to show how they feel about a topic. They write with feeling and share their opinions. Reread this passage from "Here Comes Solar Power."

Expert Model

Opinions

How does the author feel about alternative energy? What details does the author give to support the opinions?

But these fuels come from deep under the Earth's surface, and they are running out. They cannot be reused. Once a fossil fuel is gone, it's gone forever. So we need alternative energy sources to replace them.

Holger Burmeister/Alamy

Writers

Editing Marks

≡ Make a capital letter.

/ Make a small letter.

⊙ Add a period.

∧ Add.

9 Take out.

Martin wrote about why electricity is important. Read his revision.

Grammar Handbook

Pronoun-Verb Contractions
See page 490.

Student Model

WE NEED ELECTRICITY!

I think electricity is important.
∧Think about a world without it!

we would not have electric lights
≡

or stoves. We ᵈw~~would~~ be lost without
∧

our TVs and computers. Electricity

runs all of those things⊙ It also
∧

runs almost everything we use at

home. Id be sad if we didn't have

electricity.

Your Turn

COLLABORATE

- ☑ Identify an opinion and details that support It.
- ☑ Identify pronoun-verb contractions.
- ☑ Tell how revisions improved the writing.

Go Digital!
Write online in Writer's Workspace

Think It Over

Diamond in the Rough

Dad called it a diamond in the rough,
A big, old field outside of town
Now just a dumping ground
A ton of trash
      A mile of mess
"No way!" my friends say.

But I just shake my head and smile.
I know.

I can clean it. Make it green.
Polish up that diamond in the rough.

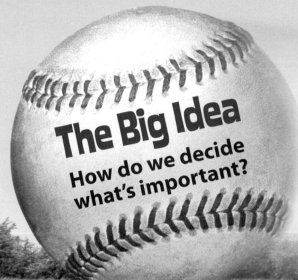

The Big Idea

How do we decide what's important?

I go each day,
Pick up trash,
Haul away sack after sack
        After sack.

My friends follow,
        Watching, watching,
Then roll up their sleeves.
They see it too—
        A diamond in the rough.

Together we yank up weeds,
        Rake things smooth,

Until finally, finally
        A baseball field.

Not the biggest
Not the best

But OURS
        A diamond in the rough.

                — By Maureen Wong

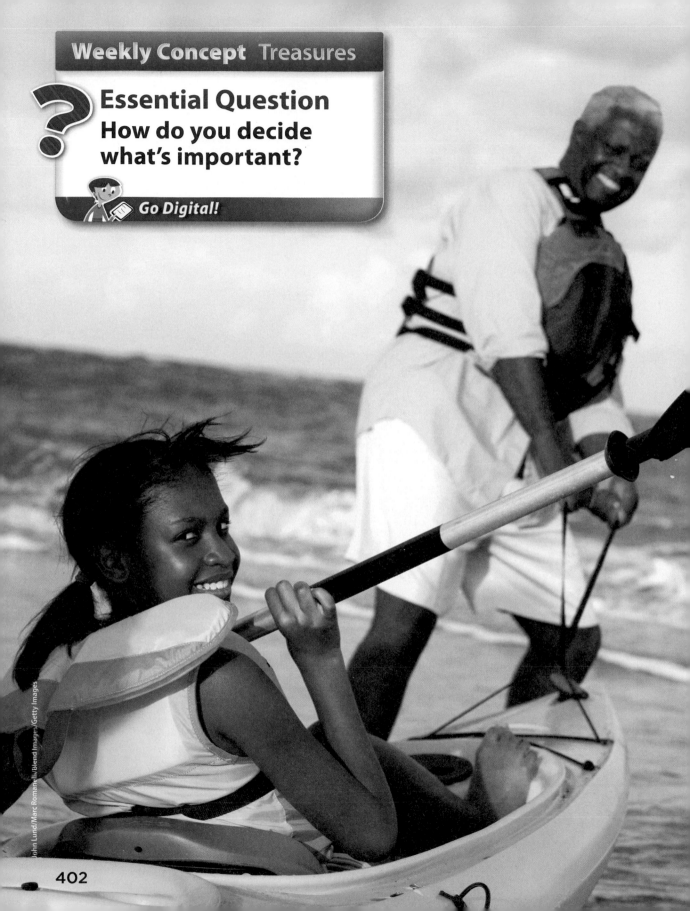

? Essential Question
How do you decide what's important?

Go Digital!

WHAT WE VALUE

Spending time with my grandfather is important. He shares what he knows and helps me learn new things. He values spending time with me, too.

▶ Spending time with people and learning new things are valuable to me.

▶ Healthy habits and sharing are important, too.

Talk About It

Write words you have learned about what's valuable. Talk with a partner about the things that you value.

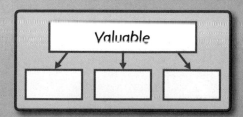

Valuable

Vocabulary

Use the picture and the sentence to talk with a partner about each word.

alarmed

Jess was **alarmed** as he watched the barber cut his hair.

Show how you would look if you were alarmed by something.

anguish

Andy felt **anguish** when he realized his bike was missing.

What is another word that means the same as anguish?

necessary

Food is **necessary** for all living things.

What other things are necessary for living things?

obsessed

Paul is **obsessed** with space and wears his astronaut suit all the time.

Name something you are obsessed with.

possess

Dan and Meg **possess** a huge bunch of colorful balloons.

Tell about something you possess.

reward

Dad took us on vacation as a **reward** for getting good grades in school.

What reward would you like to get?

treasure

Lila found a real **treasure** at the book sale.

Tell about a treasure you have.

wealth

We are counting our money and will share our **wealth** by donating it.

What is another word for *wealth*?

Your Turn

COLLABORATE

Pick three words. Write three questions for your partner to answer.

Go Digital! **Use the online visual glossary**

ATHENA AND ARACHNE

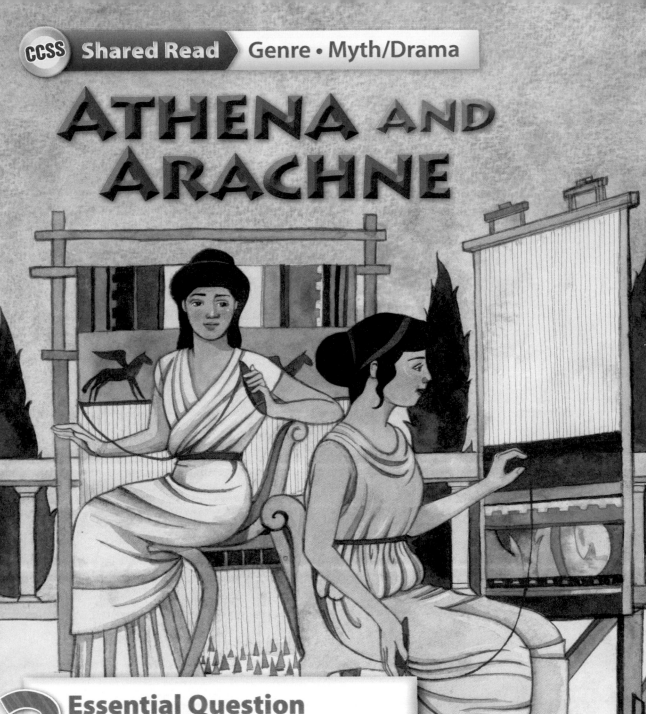

Jenny Reynish

? Essential Question

How do you decide what's important?

Read a myth that shows why valuing a talent can cause problems.

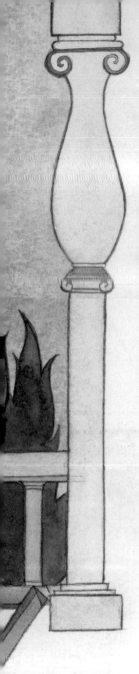

CHARACTERS

NARRATOR

ARACHNE: (uh-RAK-nee) a weaver

DIANA: Arachne's friend

ATHENA: a Greek goddess

MESSENGER

⇒ SCENE ONE ⇐

Athens, Greece, a long time ago, Arachne's home.

NARRATOR: Long ago, Arachne and her friend Diana sat weaving.

DIANA: Oh, Arachne! That cloth is so beautiful.

Arachne admires her cloth.

ARACHNE: I know. Many people want to **possess** my cloth, but few can afford it. Only those with great **wealth** can buy it.

DIANA: Yes, it's true that people value your cloth. It is one of their most valued possessions. Your weavings are a real **treasure**. Some say that you learned your weaving skill, or talent, from the goddess Athena.

ARACHNE: It was not **necessary** for me to learn from a goddess. I was born with my talent. I am a much better weaver than Athena, and I'm sure I could beat her in a weaving competition!

Diana is worried, stops weaving and looks at Arachne.

DIANA: Ssshhh! I hope Athena isn't listening, or you're in big trouble!

ARACHNE: Nonsense! There's no reason to be **alarmed** or worried. Athena is much too busy to come down from Mount Olympus to compete with me.

SCENE TWO

Mount Olympus, home of Athena. A messenger arrives.

MESSENGER: Goddess Athena! I have news from Athens. The weaver Arachne says she can beat you in a weaving competition. She's **obsessed** with her skill and thinks she is the best weaver in Greece!

ATHENA: I'll show her who weaves the finest cloth! Her obsession with weaving must end. Please get me my cloak. *Messenger hands Athena her cloak.*

ATHENA: Arachne cannot talk about me that way! If she refuses to apologize, I will make her pay for her boastful words. Her **anguish** will be great!

SCENE THREE

Arachne's home. There is a knock at the door.

ARACHNE: Who's there?

ATHENA: Just an old woman with a question.

Athena is hiding under her cloak. She enters the room.

ATHENA: Is it true that you challenged the goddess Athena to a weaving competition?

ARACHNE: Yes, that's right. *Athena drops her cloak.*

ATHENA: Well, I am Athena, and I am here to compete with you!

DIANA: Arachne, please don't! It is unwise to compete with a goddess!

Arachne and Athena sit down at the empty looms and begin to weave furiously.

ARACHNE: I am ready to win and get my **reward**!

ATHENA: There's no prize if you lose!

NARRATOR: Arachne and Athena both wove beautiful cloths. However, Arachne's cloth was filled with pictures of the gods being unkind.

ATHENA: Arachne, your weaving is beautiful, but I am insulted and upset by the pictures you chose to weave. You are boastful, and your cloth is mean and unkind. For that, I will punish you.

Athena points dramatically at Arachne. Arachne falls behind her loom and crawls out as a spider.

ATHENA: Arachne, you will spend the rest of your life weaving and living in your own web.

NARRATOR: Arachne was mean and boastful, so Athena turned her into a spider. That's why spiders are now called arachnids. Arachne learned that bragging and too much pride can lead to trouble.

⊰≡ **THE END** ≡⊱

Make Connections

What does Arachne value? How does it cause her trouble? **ESSENTIAL QUESTION**

What do you value? Why do you value it? **TEXT TO SELF**

Make Predictions

Use details in the story to predict what happens next. Was your prediction right? Read on to check it. Change your prediction if it is not right.

 Find Text Evidence

You may have made a prediction about Arachne at the beginning of "Athena and Arachne." What clues on page 407 helped you guess what might happen?

> **page 407**
>
> **DIANA:** Oh, Arachne! That cloth is so beautiful.
> *Arachne admires her cloth.*
> **ARACHNE:** I know. Many people want to **possess** my cloth, but few can afford it. Only those with great **wealth** can buy it.
> **DIANA:** Yes, it's true that people value your cloth. It is one of their most valued possessions. Your weavings are a real **treasure**. Some say that you learned your weaving skill, or talent, from the goddess Athena.
> **ARACHNE:** It was not **necessary** for me to learn from a goddess. I was born with my talent. I am a much better weaver than Athena, and I'm sure I could beat her in a weaving competition!

I predicted that Arachne and Athena would compete. I read that Arachne says she is a better weaver than Athena and could beat her in a contest. *I will read on to check my prediction.*

Your Turn

COLLABORATE

What did you predict would happen when Athena went to see Arachne? Reread page 408 to check your prediction. Remember to make, confirm, and revise predictions as you read.

Jenny Reynish

Theme

The theme of a story is the author's message. Think about what the characters do and say. This will help you figure out the theme.

 Find Text Evidence

In "Athena and Arachne," Arachne learns that bragging and too much pride can lead to trouble. This is the story's theme. I can reread to find details that help me figure out the theme.

Detail

Arachne said that many people want to possess her cloth, but few can afford it. Only those with great wealth can buy it.

↓

Detail

↓

Detail

↓

Detail

↓

Theme

Bragging and too much pride can lead to trouble.

Your Turn

COLLABORATE

Read "Athena and Arachne." List important details about what Arachne says and does in your graphic organizer. Be sure the details tell about the theme.

Go Digital!
Use the interactive graphic organizer

Myth/Drama

"Athena and Arachne" is a myth and a drama, or play. A **myth** tells how something came to be. A **drama:**

- Tells a story through dialogue and is performed
- Is separated into scenes and has stage directions

Find Text Evidence

I see that "Athena and Arachne" is a myth and a play. It is divided into three scenes. It uses dialogue and stage directions to tell how spiders came to weave webs.

page 408

⟫⟫ SCENE TWO ⟪⟪

Mount Olympus, home of Athena. A messenger arrives.

MESSENGER: Goddess Athena! I have news from Athens. The weaver Arachne says she can beat you in a weaving competition. She's **obsessed** with her skill and thinks she is the best weaver in Greece!

ATHENA: I'll show her who weaves the finest cloth! Her obsession with weaving must end. Please get me my cloak. *Messenger hands Athena her cloak.*

ATHENA: Arachne cannot talk about me that way! If she refuses to apologize, I will make her pay for her boastful words. Her **anguish** will be great!

⟫⟫ SCENE THREE ⟪⟪

Arachne's home. There is a knock at the door.

ARACHNE: Who's there?

ATHENA: Just an old woman with a question.

Athena is hiding under her cloak. She enters the room.

ATHENA: Is it true that you challenged the goddess Athena to a weaving competition?

ARACHNE: Yes, that's right. *Athena drops her cloak.*

ATHENA: Well, I am Athena, and I am here to compete with you!

Scene A scene is a part of a play. Scenes tell the story in time order.

Stage Directions Stage directions tell what the characters do and how they move.

Dialogue Dialogue is the words the characters speak.

COLLABORATE

Your Turn

Find more examples of scenes, dialogue, and stage directions in "Athena and Arachne." Tell your partner how they help tell the story.

Root Words

A root word is the simplest form of a word. It helps you figure out the meaning of a related word.

 Find Text Evidence

In "Athena and Arachne," I see the word competition. *I think the root word of* competition *is* compete. *I know* compete *means "to be in a contest." I think a* competition *is "a contest where people try to win."*

I am a much better weaver than Athena, and I'm sure I could beat her in a weaving competition!

Your Turn

Find the root word. Then use it to figure out the meaning of each word.

possessions, *page 407*

obsession, *page 408*

Jenny Reynish

Readers to ...

Writers use short and long sentences to add rhythm to what they write. Combining sentences of different lengths helps make writing more interesting.

Sentence Length

Find short and long **sentences.** Why do writers use different sentence lengths in plays?

Expert Model

MESSENGER: Goddess Athena! I have news from Athens. The weaver Arachne says she can beat you in a weaving competition. She's obsessed with her skill and thinks she is the best weaver in Greece!

ATHENA: I'll show her who weaves the finest cloth! Her obsession with weaving must end. Please get me my cloak.

Writers

Cassie wrote about something she values and why. Read her revisions.

Editing Marks

☰ Make a capital letter.

╱ Make a small letter.

⊙ Add a period.

∧ Add

ዖ Take out.

Grammar Handbook

Adjectives and Articles
See page 491.

Student Model

MR. BEAR

, very old, brown,
He is soft ∧ and cuddly. His name

is Mr. Bear. My grandmother gave

                    a
him to me. Mr. Bear is ~~the~~ little

        he
shabby, but ~~his~~ reminds me of how

much my grandmother loves me ⊙

    Because I like Mr. Bear so

much, I am writing a story about

a funny adventure story
him. It is ∧ about a stuffed bear who

comes to life and helps people.

Your Turn COLLABORATE

☑ Identify long and short sentences.
☑ Identify articles and adjectives.
☑ Tell how revisions improved the writing.

Go Digital!
Write online in Writer's Workspace

Essential Question
How can weather affect us?

Go Digital!

Ryan McVay/Lifesize/Getty Images

416

WEATHER AFFECTS US

How's the weather? Turn on the weather report or look outdoors to find out. Is it sunny? Is it raining? It's important to know what the weather is. What's today's forecast? Warm and rainy with a side of fun.

▶ Weather is what is happening outdoors right now.

▶ It affects how we dress and what we do.

▶ Sometimes severe weather affects the way we live.

Talk About It

Write words you have learned about weather. Talk with a partner about how weather affects you.

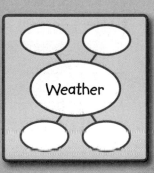

Weather

Vocabulary

Use the picture and the sentence to talk with a partner about each word.

argue

James and his grandfather sometimes discuss movies and **argue** about them.

What are some words that mean the opposite of argue?

astonished

Tammy and Kim were **astonished** and amazed at the big bug they found.

What is another word for astonished?

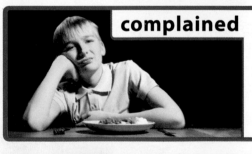

complained

Ben **complained** about his dinner because he didn't like peas.

Tell about a time when you complained about something.

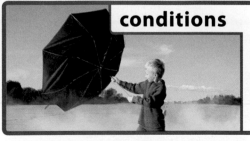

conditions

Mark's umbrella was ruined by the windy **conditions**.

What are the weather conditions today?

forbidding

This sign is **forbidding** us to swim in the pond.

Think of some words that mean the same as forbidding?

forecast

The weatherman says that the **forecast** today is sunny and warm.

Why do you look at a weather forecast?

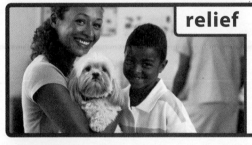

relief

It was a **relief** that Mom found my dog, and he was okay.

Tell about a time when you felt relief.

stranded

Patricia was **stranded** at home during the rainstorm.

Why might you be stranded in bad weather?

Your Turn

COLLABORATE

Pick three words. Write three questions for your partner to answer.

Go Digital! *Use the online visual glossary*

The Big Blizzard

Essential Question

How can weather affect us?

Read how a blizzard affects the Hernandez family in New York City.

Rosa and Eddie Hernandez huddled close to the radio and listened carefully to the scratchy voice of the news announcer.

"The blizzard of 1947 is the biggest snowstorm in New York City history! Tremendous amounts of snow and terrible weather **conditions** caused the city's subway system to shut down yesterday from Wall Street to Spanish Harlem. Parents are even **forbidding** their children to go outside because it is so dangerous. The weather **forecast** for today predicts that the snow will stop. In the meantime, Mayor O'Dwyer's message to all New Yorkers is this: Help each other in the face of this disaster."

"Oh, mamá!" whispered Rosa. "Will papá ever get home from work?"

Mamá gave Rosa a big hug. "He must be stuck at work and unable to get home," she said. "He is **stranded**, but don't worry. The snow is slowing down now, and I'm sure he will make it home soon."

Mamá went into the kitchen to make lunch. She came out carrying her coat and scarf.

"We are out of milk and bread, so I need to try to get to Maria's Market," said mamá.

Rosa and Eddie jumped up and begged to go with her. Mamá had kept them inside for two days because it was snowing too hard to go out.

"No," said mamá. "It's too cold."

Rosa and Eddie knew they shouldn't **argue** with mamá, but they were tired of being indoors.

"Oh please, take us outside! We can all go to the store together!" said Eddie.

"Okay," said mamá with a sigh. "But we have to stick together and stay close to each other."

Mamá helped Rosa and Eddie bundle up in their uncomfortable, but warm, wool clothes. When they got outside, they were **astonished** and amazed to find a wall of snow several feet high! Luckily, their neighbor Mr. Colón arrived with two metal shovels.

"Who wants to help dig out?" he asked.

Mamá, Rosa, and Eddie took turns shoveling snow. It was hard work, but no one fussed or **complained**. When they were done, they looked across the street. Maria's Market was still snowed in. Mrs. Sanchez, the owner, was trying to clear the snow with a small broom.

"Mr. Colón, may we borrow your shovels, *por favor*?" asked Rosa. "I think we need to give Mrs. Sanchez a hand."

Shoveling the walk in front of the store was easy. It was a piece of cake for Rosa and Eddie. They laughed and threw snowballs, too. Mrs. Sanchez was grateful for their help. "*Gracias*," she said, and gave mamá milk and bread from her store as thanks.

As Rosa and Eddie crossed the snowy street with mamá to go home, they heard a deep, familiar voice.

"Is that my Rosa and Eddie?"

"Papá!" they shouted and ran over to him. Rosa told him breathlessly about how they helped Mr. Colón and Mrs. Sanchez.

"It is such a **relief** and a comfort to finally be home," said papá. "I am so proud of you for helping our neighbors."

Make Connections

How does the weather affect the Hernandez family? **ESSENTIAL QUESTION**

Tell about a time when you or your family helped out in bad weather. **TEXT TO SELF**

Stacey Schuett

Make Predictions

Use details in the story to tell, or predict, what happens next. Was your prediction right? Read on to check it. Change it if it is not right.

 Find Text Evidence

You may have made a prediction about *papá* as you read "The Big Blizzard." What details on page 421 helped you tell what might happen?

> **page 421**
>
> Street to Spanish Harlem. Parents are even **forbidding** their children to go outside because it is so dangerous. The weather **forecast** for today predicts that the snow will stop. In the meantime, Mayor O'Dwyer's message to all New Yorkers is this: Help each other in the face of this disaster."
>
> "Oh, mamá!" whispered Rosa. "Will papá ever get home from work?"
>
> Mamá gave Rosa a big hug. "He must be stuck at work and unable to get home," she said. "He is **stranded**, but don't worry. The snow is slowing down now, and I'm sure he will make it home soon."
>
> Mamá went into the kitchen to make lunch. She came out carrying her coat and scarf.
>
> "We are out of milk and bread, so I need to try to get

I predicted that papá would come home soon. I read that <u>the radio announcer said the snow would stop. Mamá told Rosa that she was sure he would make it home soon.</u> I will read on to check my prediction.

Your Turn

COLLABORATE

What did you predict would happen when Mr. Colón arrived with two metal shovels? Read page 422. Was your prediction correct? Did it need to be revised?

Theme

The theme of a story is the author's message. Think about what the characters do and say. This will help you figure out the theme.

Find Text Evidence

In "The Big Blizzard," Mayor O'Dwyer tells New Yorkers to help one another. I think this is an important detail about the theme. I will read to find more details about the characters' actions. Then I can figure out the story's theme.

Detail
Mayor O'Dwyer's message to all New Yorkers is, "Help one another in the face of this disaster."

↓

Detail
The Hernandez family helps Mr. Colón shovel snow.

↓

Detail

↓

Detail

↓

Theme

Your Turn

Reread "The Big Blizzard." Find more important details and write them in your graphic organizer. Then use the details to figure out the theme.

Go Digital!
Use the interactive graphic organizer

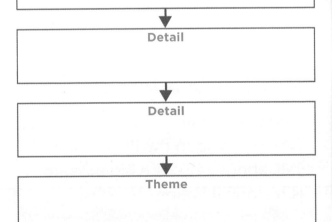

Historical Fiction

"The Big Blizzard" is historical fiction. **Historical fiction**:
- Is a made-up story that takes place in the past
- Has illustrations that show the setting and how people lived, and often helps create a mood

Find Text Evidence

I can tell that "The Big Blizzard" is historical fiction. The characters and events are made up. The story is based on real events that happened in 1947 in New York City.

page 422

Rosa and Eddie knew they shouldn't **argue** with mamá, but they were tired of being indoors.

"Oh please, take us outside! We can all go to the store together!" said Eddie.

"Okay," said mamá with a sigh. "But we have to stick together and stay close to each other."

Mamá helped Rosa and Eddie bundle up in their uncomfortable, but warm, wool clothes. When they got outside, they were **astonished** and amazed to find a wall of snow several feet high! Luckily, their neighbor Mr. Colón arrived with two metal shovels.

"Who wants to help dig out?" he asked.

Mamá, Rosa, and Eddie took turns shoveling snow. It was hard work, but no one fussed or **complained**. When they were done, they looked across the street. Maria's Market was still snowed in. Mrs. Sanchez, the owner, was trying to clear the snow with a small broom.

"Mr. Colón, may we borrow your shovels, *por favor*?" asked Rosa. "I think we need to give Mrs. Sanchez a hand."

422

The story and characters are made up, but the events could happen in real life. Events in historical fiction happened a long time ago.

Illustrations show details of the setting and how people lived.

Your Turn
COLLABORATE

Find two details about the past in the story and in the illustrations. Talk about why "The Big Blizzard" is historical fiction.

Idioms

An idiom is a group of words that mean something different from the meaning of each word in it. The phrase *under the weather* is an idiom. It doesn't mean someone is outside in bad weather. It means that someone feels sick.

 Find Text Evidence

On page 422, the phrase stick together *is an idiom. I can use clues in the story to help me figure out that it means "to stay close together in a group."*

> Oh please, take us outside! We can all go to the store together!" said Eddie.
>
> "Okay," said mamá." But we have to <u>stick together</u> and stay close to each other."

Your Turn

Talk about these idioms from "The Big Blizzard."
bundle up, *page 422*
give Mrs. Sanchez a hand, *page 422*
a piece of cake, *page 423*

Stacey Schuett

427

Readers to...

Writers use linking words to connect ideas and to show how ideas are the same or different. Some examples of linking words are *and, another, also, but,* and *more.* Reread the passage from "The Big Blizzard."

Linking Words

Where does the writer use **linking words** to connect or compare ideas?

Expert Model

Mamá gave Rosa a big hug. "He must be stuck at work and unable to get home," said mamá. "He is stranded, but don't worry. The snow is slowing down now, and I'm sure he will make it home soon."

Mamá went into the kitchen to make lunch. She came out carrying her coat and scarf.

"We are out of milk and bread, so I need to try to get to Maria's Market," said mamá.

Stacey Schuett

Writers

Maggie wrote about her experience with a hurricane. Read her revisions.

Student Model

Hurricane Night

The hurricane wasn't supposed
to reach our town until morning, but it
came in the middle of the night.

the strong wind and rain hit my

window and woke me up. It was the
                    est
stronger wind and rain I had ever
        However,
seen. I wasn't a bit scared because

I knew I was safe in my house.

My mother and brothers were
                  .
also close by. They helped me
      brave
feel bravest.

Editing Marks

≡ Make a capital letter.

/ Make a small letter.

⊙ Add a period.

∧ Add.

Take out.

Grammar Handbook

Adjectives that Compare
See page 492.

Your Turn

☑ Identify linking words.
☑ Identify adjectives that compare.
☑ Tell how revisions improved the writing.

Go Digital!
Write online in Writer's Workspace

Essential Question
Why are goals important?

Go Digital!

SUCCESS!

Kayla had a goal. She wanted to win her race in the Special Olympics. Her goal was important to her, so she worked hard. Kayla is so proud of herself today.

▶ Goals are important.

▶ They help us focus and learn new things.

▶ Achieving our goals makes us feel good about ourselves.

Talk About It

Write words you have learned about goals. Talk with a partner about why goals are important.

Vocabulary

Use the picture and the sentence to talk with a partner about each word.

communicated

Mora and her friends **communicated** by writing emails to each other.

What are some ways you have communicated with friends?

essential

A toothbrush is an **essential** tool for cleaning teeth.

What is an essential tool for planting seeds?

goal

Nick reached his **goal** and learned to swim.

Tell about a goal that you have.

motivated

Jerry was **motivated** to learn to play his guitar, so he practiced every day.

What have you been motivated to learn to do?

professional

Ted works as a **professional** police officer.

Where would a professional musician work?

research

Melanie's mom is a scientist, and she uses a microscope to do **research**.

What animal would you like to research?

serious

Winnie pays attention because she is **serious** about getting good grades.

What is another word for *serious*?

specialist

Dr. Morrison is a **specialist** in sports medicine.

What does it take to be a specialist?

Your Turn

COLLABORATE

Pick three words. Write three questions for your partner to answer.

Go Digital! *Use the online visual glossary*

ROCKETING INTO SPACE

Essential Question

Why are goals important?

Read how one man used his education and experience to reach his goals.

When James A. Lovell, Jr. was a boy, he loved to build rockets and launch them into the sky. But his dreams went a lot farther than his rockets. Like many boys who grew up in the 1930s, Lovell dreamed of being a pilot. And as he watched his rockets soar, he knew someday he would, too.

HIGH FLYING DREAMS

Lovell was born in Cleveland, Ohio, in 1928. He worked hard in school and planned to go to a special college to study **astronomy** and rockets. Unfortunately, he didn't have enough money to attend. Lovell had to figure out another way to reach his **goal**.

Lovell was **motivated** to find a way to fly rockets. So, he went to college near his home for two years and then signed up for flight training at the United States Naval Academy. After four years at the academy, Lovell joined the United States Navy and became a **professional** naval test pilot. His job was to fly planes before anyone else was allowed to fly them.

James A. Lovell, Jr. became an astronaut in 1962. He flew four space missions.

PILOT TO ASTRONAUT

As a pilot, Lovell spent more than half of his flying time in jets. He taught other pilots how to fly. He also worked as a **specialist** in air flight safety. Soon, the National Aeronautics and Space Administration, or NASA, put out a call for astronauts. Lovell applied for the job because he had all the **essential** skills needed to fly into space. As a result, NASA chose him. By 1962, James Lovell was an astronaut! He had finally reached his goal.

BIG CHALLENGES

Lovell flew on three space missions, and then, in April 1970 he became commander of the Apollo 13 mission. This was a big responsibility and a great honor. This was also one of the biggest **challenges** of Lovell's life.

Apollo 13 was supposed to land on the Moon. Two days after leaving Earth, however, the spacecraft had a **serious** problem. One of its oxygen tanks exploded. The crew did not have enough power or air to breathe. They could not make it to the Moon.

Lovell **communicated** with the experts at NASA. No one knew what to do at first. Then the team on the ground did some **research** and came up with a solution. The astronauts followed the team's directions and built an invention using plastic bags, cardboard, and tape. It worked! It cleaned the air in the spacecraft. But the next problem was even bigger. How were the astronauts going to get back to Earth?

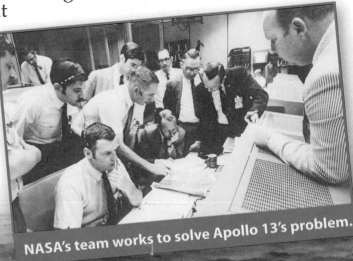

NASA's team works to solve Apollo 13's problem.

A JOB WELL DONE

The NASA team decided the astronauts would use the **lunar,** or moon, module as a lifeboat. James and the other two astronauts climbed into the smaller spacecraft and shut the hatch tight. They moved away from the main spaceship. With little power, water, food, or heat, the astronauts listened carefully to the team at NASA.

The trip back to Earth was dangerous and scary. For almost four days, the astronauts traveled in a cramped capsule. They were cold, thirsty, and hungry. Then, with millions of people watching on television, the module fell to Earth.

The Apollo 13 crew splashed down safely on April 17, 1970.

Years later, James Lovell said that Apollo 13 taught him how important it was for people to work together. His favorite memory was when the capsule splashed down in the Pacific Ocean and the diver knocked on the window to let them know they were safe.

A DREAM COME TRUE

DID YOU EVER DREAM OF GOING INTO SPACE? CHECK OUT SPACE CAMP!

Space camps have been around for more than 30 years. They make science, math, and technology exciting so kids will want to learn more. And like the NASA training programs, these camps teach the importance of teamwork and leadership.

? Make Connections

How did James A. Lovell's goals as a child help him as an adult? **ESSENTIAL QUESTION**

Tell about one of your goals and how you might achieve it. **TEXT TO SELF**

Reread

Stop and think about the text as you read. Are there new facts and ideas? Do they make sense? Reread to make sure you understand.

 Find Text Evidence

Do you understand what James A. Lovell, Jr. did to become a pilot? Reread "High Flying Dreams" on page 435.

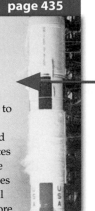

page 435

in 1928. He worked hard in school and planned to go to a special college to study **astronomy** and rockets. Unfortunately, he didn't have enough money to attend. Lovell had to figure out another way to reach his **goal**.

Lovell was **motivated** to find a way to fly rockets. So, he went to college near his home for two years and then signed up for flight training at the United States Naval Academy. After four years at the academy, Lovell joined the United States Navy and became a **professional** naval test pilot. His job was to fly planes before

I read that James Lovell <u>went to college and then to the United States Naval Academy. He signed up for flight training and became a professional naval test pilot.</u> James Lovell became a pilot by going to school. He never gave up.

Your Turn

COLLABORATE

How did James Lovell help get his Apollo 13 space ship back home? Reread pages 436 and 437.

Problem and Solution

Some informational texts describe a problem, tell the steps taken to solve the problem, and give the solution. Signal words, such as *problem, solution, solve,* and *as a result,* show there is a problem and the steps to a solution.

 ## Find Text Evidence

James Lovell wanted to fly rockets but didn't have enough money to go to a special college. That was his problem. What steps did he take to solve his problem? What was the solution?

Problem

James wanted to fly rockets, but he didn't have enough money to go to a special college.

↓

James went to college near his home.

↓

He became a test pilot.

↓

He joined NASA and became an astronaut.

↓

Solution

As an astronaut, James Lovell was able to fly in rocket ships.

Your Turn

Reread "Big Challenges." What was James Lovell's problem on Apollo 13? Find the steps he took to solve it and write them in your graphic organizer. Then write the solution.

Go Digital!
Use the interactive graphic organizer

Biography

"Rocketing Into Space" is a biography. A **biography**:
- Tells the true story of a real person's life
- Is written by another person
- Includes text features such as keywords, photographs and captions

 Find Text Evidence

I can tell that "Rocketing Into Space" is a biography. It is the true story of James Lovell's life. It has photographs with captions and key words that are important to the biography.

page 436

PILOT TO ASTRONAUT

As a pilot, Lovell spent more than half of his flying time in jets. He taught other pilots how to fly. He also worked as a **specialist** in air flight safety. Soon, the National Aeronautics and Space Administration, or NASA, put out a call for astronauts. Lovell applied for the job because he had all the **essential** skills needed to fly into space. As a result, NASA chose him. By 1962, James Lovell was an astronaut! He had finally reached his goal.

BIG CHALLENGES

Lovell flew on three space missions, and then, in April 1970 he became commander of the Apollo 13 mission. This was a big responsibility and a great honor. This was also one of the biggest **challenges** of Lovell's life.

Apollo 13 was supposed to land on the Moon. Two days after leaving Earth, however, the spacecraft had a **serious** problem. One of its oxygen tanks exploded. The crew did not have enough power or air to breathe. They could not make it to the Moon.

Lovell **communicated** with the experts at NASA. No one knew what to do at first. Then the team on the ground did some **research** and came up with a solution. The astronauts followed the team's directions and built an invention using plastic bags, cardboard, and tape. It worked! It cleaned the air in the spacecraft. But the next problem was even bigger. How were the astronauts going to get back to Earth?

NASA's team works to solve Apollo 13's problem.

436

Text Features

Keywords Keywords are important words. They are in dark type.

Photographs Photographs and their captions show more about the events in the person's life.

Your Turn COLLABORATE

Find another keyword. Why is this an important word in James Lovell's biography?

Greek and Latin Roots

Many words have word parts, such as Greek or Latin roots, in them. The Greek root *astro* means "star," and *naut* means "ship." The Latin root *luna* means "moon."

 Find Text Evidence

On page 435, I see the word astronomy. *I remember that* astro *comes from a Greek root that means "star." I think* astronomy *must have something to do with the stars. It may mean "the study of the stars."*

He worked hard in school and planned to go to a special college to study astronomy and rockets.

Your Turn

Use the Greek or Latin roots to figure out the meaning of the each word.

astronauts, *page 436*
lunar, *page 437*

NASA

Readers to...

Writers organize their ideas when they write. Putting ideas in order helps the reader make sense of the text. Reread this passage from "Rocketing Into Space."

Ideas in Order

How does the way the author put **ideas in order** help you understand the text?

Expert Model

Lovell communicated with the experts at NASA. No one knew what to do at first. Then the team on the ground did some research and came up with a solution. The astronauts followed the team's directions and built an invention using plastic bags, cardboard, and tape. It worked! It cleaned the air in the spacecraft. But the next problem was even bigger. How were the astronauts going to get back to Earth?

NASA

442

Writers

Editing Marks

= Make a capital letter.

/ Make a small letter.

⊙ Add a period.

∧ Add.

~ Take out.

Grammar Handbook

Adverbs That Tell How See page 493.

Leah wrote about why going to school is important. Read Leah's revision.

Student Model

Why School Is Important

Some people don't like to go to

school. I think it is very important.

          first
When I ∧started kindergarten, I

                                   ly
did not know how to read. I slow∧

         every day
learned new things∧. I learned

to read, write, do math, and

                   Now that
understand science. ~~Because~~ I am

in the third grade, I can do all of

          really well
that ~~good~~∧.

Your Turn COLLABORATE

✔ Identify how the writer ordered ideas.

✔ Identify adverbs that tell how.

✔ Tell how revisions improved the writing.

Go Digital!
Write online in Writer's Workspace

Essential Question
How can learning about animals help you respect them?

Go Digital!

Peter Dazeley/Photographer's Choice/Getty Images

444

Respecting Animals

My name is Max and I'm a Great Horned Owl. My job here at the wildlife preserve is to help kids like Nora learn all about owls. All animals deserve respect.

Learning about animals helps you respect them.

The more you know about animals, the more you can do to help them.

Talk About It

Write words you have learned about respecting animals. Talk with a partner about ways you can help protect animals.

Vocabulary

Use the picture and the sentence to talk with a partner about each word.

endangered

The giant panda is an **endangered** animal and needs to be protected.

Name another endangered animal.

fascinating

Maya thought the butterflies on her shirt were **fascinating** and interesting.

Tell about something you find fascinating.

illegal

The sign says it is **illegal** to swim here because the beach is closed.

What word means the opposite of illegal?

BEACH CLOSED

inhabit

Many animals, including moose, **inhabit** forests.

What animals inhabit the ocean?

requirement

Food is an important **requirement** for all living things.

Name another requirement for living things.

respected

The players **respected** their football coach because he was smart and fair.

Who is respected in your school?

unaware

The Karr family is **unaware** that there is a giraffe watching them.

What word means the opposite of unaware?

wildlife

Zebras are one kind of **wildlife** that live in Africa.

What kind of wildlife lives near your home?

Your Turn

COLLABORATE

Pick three words. Write three questions for your partner to answer.

Go Digital! *Use the online visual glossary*

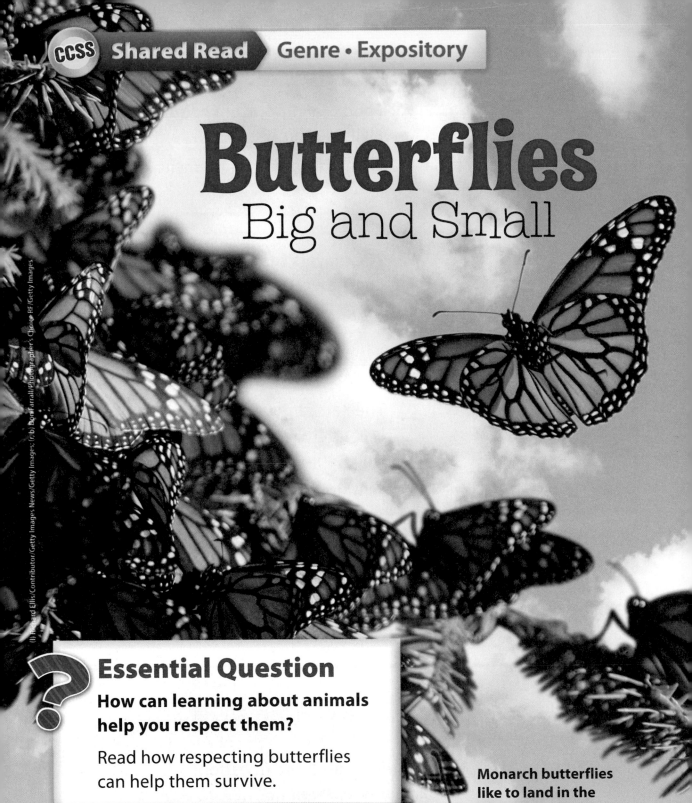

Butterflies
Big and Small

(l) Richard Ellis/Contributor/Getty Image: News/Getty Images; (r, b) Don Farall/Photographer's Choice RF/Getty Images

? Essential Question

How can learning about animals help you respect them?

Read how respecting butterflies can help them survive.

Monarch butterflies like to land in the same trees when they migrate.

There are more than 725 species, or kinds, of butterflies fluttering around the United States and Canada. These **fascinating** creatures taste leaves with their feet and only see the colors red, yellow, and green. The Monarch butterfly and the Western Pygmy Blue butterfly share these same traits, but they are also different in many ways.

Size and Color

The Western Pygmy Blue butterfly is the smallest butterfly in the world. It is just about a half-inch across from wing tip to wing tip. That's smaller than a dime!

Monarch butterflies are much bigger. They measure about four inches across.

Size is not the only way Monarchs are different from Pygmy Blues. Monarch butterflies are a bright orange color with black markings. That makes them easy to see. Pygmy Blue butterflies are mostly brown and blue, and they blend in with their surroundings. Many people walk right by Pygmy Blues, **unaware** that they are even there.

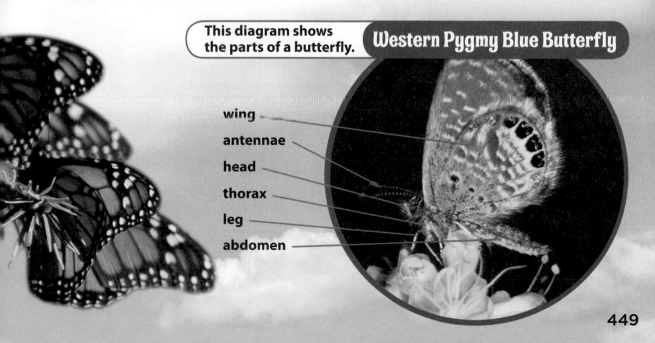

This diagram shows the parts of a butterfly.

Western Pygmy Blue Butterfly

wing
antennae
head
thorax
leg
abdomen

Moving Around

Almost all butterflies migrate, or move to different areas. The Monarch's journey is the longest migration of any butterfly in the world. It spends summers in the northern United States and Canada. Then it migrates south to Mexico in early fall. Many Monarchs travel more than 3,000 miles.

Western Pygmy Blue butterflies **inhabit** southwestern deserts and marshes from California to Texas. They migrate short distances north to Oregon, and also to Arkansas, and Nebraska.

Both Monarchs and Blue Pygmies migrate when the weather gets chilly. Butterflies are cold-blooded insects. They are hot when the weather is hot and cold when the weather is cold. As a result, both butterflies migrate to stay warm. They also journey north or south to find food.

Finding Food

The Western Pygmy Blue drinks the nectar of many kinds of flowers. It finds the sweet, thick liquid easily, so its population has steadily grown. However, Monarch butterflies are not so lucky.

This Western Pygmy Blue butterfly stops to eat.

Butterfly Migration

CANADA
Great Lakes
UNITED STATES
Pacific Ocean
MEXICO

N W E S

Map Key
← Monarch butterfly migration route
← Western Pygmy Blue butterfly migration route

Just like the Pygmy Blue, Monarch butterflies sip nectar from flowers. But the Monarch butterfly has one main food **requirement** — the milkweed. Monarch butterflies must find this plant along their migration route. But what happens if there are no milkweed leaves?

When people build houses and roads, there are fewer places for Monarchs to find milkweed. If the Monarch cannot find food, its population will decrease. The Western Pygmy Blue and Monarch butterflies are not **endangered**, or at risk for becoming extinct now, but biologists are worried. Many other butterflies are endangered because people destroy their habitats.

Help Butterflies

Like all **wildlife**, Monarch and Pygmy Blue butterflies should be **respected**. People need to preserve butterfly habitats. To help, they can work to change laws, plant milkweed, and make it **illegal** to destroy animal habitats.

Learning about butterflies and what they need to survive is important. That way there will be plenty of Western Pygmy Blue and Monarch butterflies for future generations to enjoy.

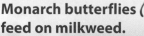

Monarch butterflies feed on milkweed.

Make Connections

How can people learn to respect butterflies? **ESSENTIAL QUESTION**

Talk about some butterflies you've seen. How are they alike and different? **TEXT TO SELF**

Reread

Stop and think about the text as you read. Are there new facts and ideas? Do they make sense? Reread to make sure you understand.

Find Text Evidence

Do you understand some ways the Monarch butterfly is different from the Western Pygmy Blue butterfly? Reread "Size and Color" on page 449.

page 449

There are more than 725 species, or kinds, of butterflies fluttering around the United States and Canada. These **fascinating** creatures taste leaves with their feet and only see the colors red, yellow, and green. The Monarch butterfly and the Western Pygmy Blue butterfly share these same traits, but they are also different in many ways.

Size and Color

The Western Pygmy Blue butterfly is the smallest butterfly in the world. It is just about a half-inch across from wing tip to wing tip. That's smaller than a dime!

Monarch butterflies are much bigger. They measure about four inches across.

Size is not the only way Monarchs are different from Pygmy Blues. Monarch butterflies are a bright orange color with black markings. That makes them easy to see. Pygmy Blue butterflies are mostly brown and blue, and they blend in with their surroundings. Many people walk right by Pygmy Blues, **unaware** that they are even there.

I read that <u>the Western Pygmy Blue butterfly is smaller than a dime and is mostly brown and blue in color. The Monarch butterfly is about four inches wide and is orange and black.</u> Now I understand some of the ways these two butterflies are different.

Your Turn

COLLABORATE

How are Monarch butterflies and Western Pygmy Blue butterflies alike? Reread "Moving Around" on page 450 to find out.

Compare and Contrast

When authors compare, they show how two things are alike. When authors contrast, they tell how the things are different. Authors use signal words, such as *both*, *alike*, *same*, or *different*, to compare and contrast.

 Find Text Evidence

How are the Monarch butterfly and Western Pygmy Blue butterfly alike and different? I will reread "Butterflies Big and Small" and look for signal words.

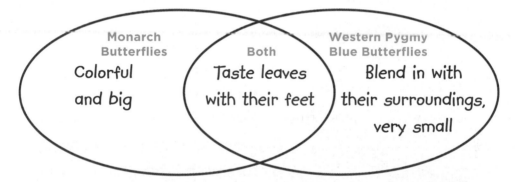

Monarch Butterflies
Colorful and big

Both
Taste leaves with their feet

Western Pygmy Blue Butterflies
Blend in with their surroundings, very small

Your Turn

Reread "Butterflies Big and Small." Find details that tell how Monarch butterflies and Western Pygmy Blue butterflies are alike and different. Write them in the graphic organizer. What signal words helped you?

Go Digital!
Use the interactive graphic organizer

453

Expository Text

"Butterflies Big and Small" is an expository text.

Expository text:

- May give information about a science topic
- Has headings that tell what a section is about
- Includes text features such as diagrams and maps

Find Text Evidence

I can tell that "Butterflies Big and Small" is expository text. It give facts about Monarch and Western Pygmy Blue butterflies. This science article also has headings, a diagram, and a map.

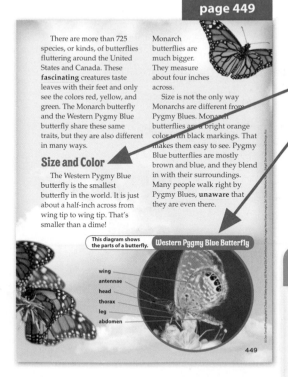

page 449

There are more than 725 species, or kinds, of butterflies fluttering around the United States and Canada. These **fascinating** creatures taste leaves with their feet and only see the colors red, yellow, and green. The Monarch butterfly and the Western Pygmy Blue butterfly share these same traits, but they are also different in many ways.

Size and Color

The Western Pygmy Blue butterfly is the smallest butterfly in the world. It is just about a half-inch across from wing tip to wing tip. That's smaller than a dime!

Monarch butterflies are much bigger. They measure about four inches across.

Size is not the only way Monarchs are different from Pygmy Blues. Monarch butterflies are a bright orange color with black markings. That makes them easy to see. Pygmy Blue butterflies are mostly brown and blue, and they blend in with their surroundings. Many people walk right by Pygmy Blues, **unaware** that they are even there.

This diagram shows the parts of a butterfly.

Western Pygmy Blue Butterfly

wing
antennae
head
thorax
leg
abdomen

449

Text Features

Headings Headings tell what a section of text is mostly about.

Diagram A diagram is a simple picture with labels.

COLLABORATE

Your Turn

Look at the text features in "Butterflies Big and Small." Tell your partner about something you learned.

Context Clues

Context clues are words or phrases that help you figure out the meaning of an unfamiliar word. In many science texts, context clues appear in the same paragraph as an unfamiliar word.

 Find Text Evidence

On page 450, I'm not sure what the word migrate *means. I will look for clues in the paragraph. I read that butterflies "move to different areas" and "travel more than 3,000 miles." I also see the word "journey." I think* migrate *means to move or travel to different places.*

> Almost all butterflies migrate, or move to different areas. The Monarch's journey is the longest migration of any butterfly in the world. It spends summers in the northern United States and Canada. Then it migrates south to Mexico in early fall. Many Monarchs travel more than 3,000 miles.

Your Turn

COLLABORATE

Find context clues. Use them to figure out the meanings of the following words.

cold-blooded, *page 450*

nectar, *page 451*

Don Farrall/Photographer's Choice RF/Getty Images

Readers to...

Writers use a strong conclusion in nonfiction writing. A strong conclusion retells the main idea in different words. Reread the passage from "Butterflies Big and Small."

Strong Conclusions

What information is in this strong **conclusion**?

Expert Model

Like all wildlife, Monarch and Pygmy Blue butterflies need to be respected. People need to preserve butterfly habitats. To help, they can work to change laws, plant milkweed, and make it illegal to destroy animal habitats.

Learning about butterflies and what they need to survive is important. That way there will be plenty of Western Pygmy Blue and Monarch butterflies for future generations to enjoy.

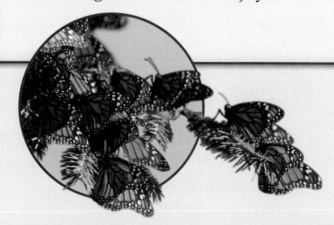

Writers

Editing Marks

≡ Make a capital letter.

∕ Make a small letter.

⊙ Add a period.

∧ Add

⌒ Take out.

Grammar Handbook

Adverbs That Compare See page 494.

Ramona wrote about earthworms and how her feelings about them changed. Read Ramona's revision.

Student Model

Earthworms

I used to be afraid of earthworms. Then I learned more about them⊙ Earthworms help plants grow ^better best⌒ they make tunnels in dirt that hold water and air. Earthworms are really the ^most more⌒ slimy of all worms. But being slimy helps them breathe. Now that I know more, I am not afraid of t̶h̶e̶m̶ ^earthworms anymore.

Your Turn

☑ Identify the strong conclusion.

☑ Identify adverbs that compare.

☑ Tell how revisions improved the writing.

Go Digital!
Write online in Writer's Workspace

Essential Question
What makes you laugh?

Go Digital!

LET'S LAUGH

Lots of things make you laugh - jokes, funny stories, silly pictures. Just look at these pink pigs. Aren't they funny?

- ▶ Having a good sense of humor is important.
- ▶ Laughing makes you feel good.
- ▶ Laughing helps you share feelings with friends.

Talk About It

Talk with a partner about what you think is humorous. Write words about things that make you laugh.

Humorous

459

Vocabulary

Use the picture and the sentence to talk with a partner about each word.

entertainment

Grandpa and Devon think playing chess is great **entertainment**.

What do you like to do for entertainment?

humorous

Evan couldn't stop laughing at Nick's **humorous** story.

Tell a humorous story to a partner.

ridiculous

Jess wore a **ridiculous** clown nose and made his friends giggle.

What is another word for ridiculous?

slithered

The long, thin snake **slithered** across the floor.

Move your hands to show what slithered looks like.

Poetry Words

narrative poem

My favorite **narrative poem** tells about Paul Revere's ride.

What story would you tell in a narrative poem?

rhyme

The words *moon* and *spoon* **rhyme** because they end in the same sound.

Name another word that rhymes with *moon* and *spoon*.

rhythm

Ben's poem has a **rhythm** that sounds like a drumbeat.

Why might a poet include rhythm in a poem?

stanza

Each **stanza** in Maggie's poem has five lines.

Write a poem with two stanzas.

COLLABORATE

Your Turn

Pick three words. Then write three questions for your partner to answer.

Go Digital! *Use the online visual glossary*

The Camping Trip

We roughed it at Old Piney Park,
With tents and hot dogs after dark.

I'd barely yawned and gone to sleep,
When I felt something creep, creep, creep.

A slimy something crawled on me,
Across my toe, up to my knee.

? Essential Question

What makes you laugh?

Read two poems about funny
situations.

Ridiculous! Hard to believe,
That creature **slithered** up my sleeve.

It was not **humorous** or fun.
I hollered "Rattlesnake! Let's run!"

We all jumped up and stomped around,
Our tent collapsed flat on the ground.

Ten flashlights clicked on to reveal,
That creepy crawly by my heel.

I blushed bright red, "Oops, I was wrong."
Snake?

      No, a lizard—one-inch long.

— **Constance Andrea Keremes**

Bubble Gum

I bought a pack of bubble gum,
    As I do every week,
Unwrapping 10 or 20 sticks,
    I popped them in my cheek.

I started masticating,
    That's a fancy word for chew,
The gum became a juicy gob,
    I took a breath and blew.

I suddenly inflated,
    Puffing up like a balloon,
I was a giant bubble,
    Big and round as a full moon.

My father hit the ceiling,
    He was really in a stew,
He hollered, "Stop! Don't go!"
    As out the door I flew.

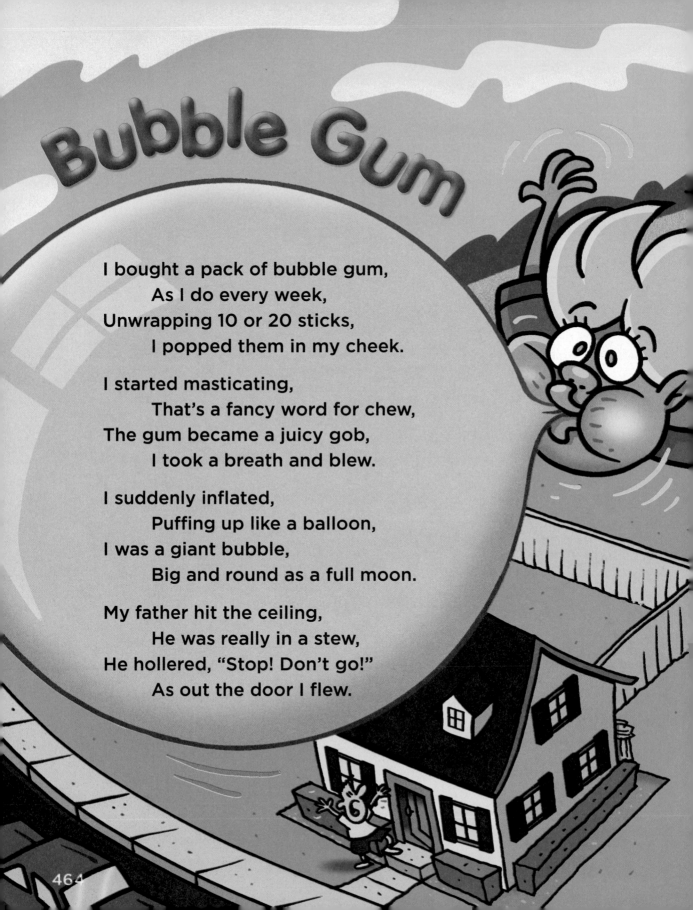

The neighbors' eyes were popping.
    They dropped everything to see.
I was the **entertainment** of the day.
    Forget about TV.

If you like bubble gum, beware—
    Chew just one stick a day,
Or you'll become a bubble, too
    And float up   Up   AWAY!

I saw my friends below me,
    And let loose a mighty roar.
WHOOSH!
All my air blew out,
    And I was just a kid once more.

— **Diana Kent**

Make Connections

Which poem made you laugh? Talk about what funny thing happens in each of the poems. **ESSENTIAL QUESTION**

Which poem has the funniest events or characters? **TEXT TO SELF**

Daryll Collins

Narrative Poem

Narrative poetry: • Tells a story. • Can have any number of lines and stanzas.

A **stanza**: • Is a group of lines that form part of a poem.
• Often has rhyme and rhythm.

Find Text Evidence

I can tell that "Bubble Gum" is a narrative poem. It tells a story. It also has stanzas. Each stanza has four lines. The second and fourth lines rhyme.

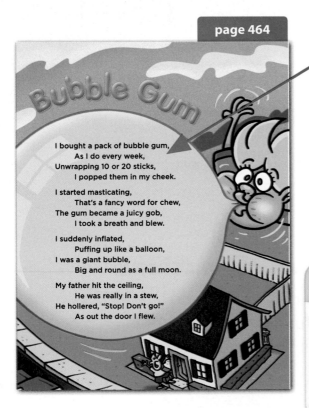

page 464

Bubble Gum

I bought a pack of bubble gum,
    As I do every week,
Unwrapping 10 or 20 sticks,
    I popped them in my cheek.

I started masticating,
    That's a fancy word for chew,
The gum became a juicy gob,
    I took a breath and blew.

I suddenly inflated,
    Puffing up like a balloon,
I was a giant bubble,
    Big and round as a full moon.

My father hit the ceiling,
    He was really in a stew,
He hollered, "Stop! Don't go!"
    As out the door I flew.

This is a stanza. It is a group of lines. There are four stanzas on this page.

Your Turn

COLLABORATE

Reread the poem "The Camping Trip." Explain why it is a narrative poem. Tell how many stanzas are in it.

Point of View

Point of view In a poem Is what the narrator thinks about an event, a thing, or a person. Look for details that show the narrator's point of view.

 Find Text Evidence

In "The Camping Trip" I read that the narrator feels something creeping on him. He calls it slimy. He says it slithered. The details tell me he is either afraid of or dislikes small, creepy crawly things. This is the narrator's point of view.

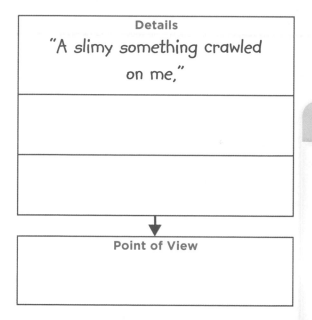

Details
"A slimy something crawled on me,"

↓

Point of View

 COLLABORATE

Your Turn

Reread "The Camping Trip." Find more details about how the narrator feels about the creepy creature. Write them in the graphic organizer. Then write the narrator's point of view. Do you agree with the narrator's point of view? Why or why not?

Go Digital!
Use the interactive graphic organizer

Rhythm and Rhyme

Poets use rhythm and rhyme make a poem interesting to listen to and fun to read.

 ## Find Text Evidence

Reread the poem "Bubble Gum" on pages 464–465 aloud. Listen for words that rhyme. Clap your hands as you read the poem to follow the poem's rhythm.

page 464

I bought a pack of bubble gum,
    As I do every week,
Unwrapping 10 or 20 sticks,
    I popped them in my cheek.

I started masticating,
    That's a fancy word for chew,

In the second and fourth lines of the poem, the words week *and* cheek *rhyme. I clapped my hands to find the rhythm. I like the way the poem has a pattern of sounds that repeat themselves.*

Your Turn

 COLLABORATE

Find more examples of rhythm and rhyme in "The Camping Trip" on pages 462–463.

Daryll Collins

Idioms

An idiom is a group of words that means something different from the usual meaning of each word in it. The phrase *lend a hand* is an idiom. It doesn't mean "to give someone your hand." It means "to help someone do something."

 Find Text Evidence

On page 462 in "The Camping Trip," the phrase roughed it *is an idiom. I can use clues in the poem to help me figure out that it means "to live without the usual comforts of home."*

page 462

We *roughed it* at Old Piney Park,
With tents and hot dogs after dark.

Your Turn

COLLABORATE

Talk about these idioms from "Bubble Gum."
hit the ceiling, *page 464*
eyes were popping, *page 465*

Daryl Collins

Readers to . . .

Writers use descriptive words to make their writing interesting and clear. Strong, or precise, words show, rather than tell. Reread the following stanzas from "Bubble Gum."

Precise Words

Identify the **strong nouns and verbs**. How do they help you visualize what the narrator is doing?

Expert Model

I started masticating,
    That's a fancy word for chew,
The gum became a juicy gob,
    I took a breath and blew.

I suddenly inflated,
    Puffing up like a balloon,
I was a giant bubble,
    Big and round as a full moon.

Daryll Collins

Writers

Ryan wrote about something that makes him laugh. Read his revisions.

Editing Marks

= Make a capital letter.

/ Make a small letter.

⊙ Add a period.

∧ Add

⌐ Take out.

Grammar Handbook

Prepositions

See page 495.

Student Model

My Dog Buddy

Buddy is the funniest dog ~~at~~ _in_ the world. When I take him for a walk after school, he stops short. Sometimes I stumble over him. That makes me _laugh_ ~~laughing~~.

Sometimes he runs so fast that he _trips_ ~~falls~~ over his own ears. And once, buddy followed me onto the school bus! He made everyone _giggle_ ~~laugh~~.

Buddy is the _silliest_ ~~sillier~~ dog I know!

Your Turn

COLLABORATE

- ✔ Identify strong nouns and verbs.
- ✔ Identify prepositions.
- ✔ Tell how revisions improved the writing.

Go Digital!
Write online in Writer's Workspace

471

Contents

Sentences

Sentences and Fragments

A **complete sentence** is a group of words that tells a complete thought. A **sentence fragment** is a group of words that does not tell a complete thought.

The rocket. (does not tell a complete thought)
The rocket soars into the clouds. (tells a complete thought)

Your Turn **Write each group of words. Tell whether it is "complete" or "fragment." Write it.**

1. My little brother.
2. My father takes a picture.
3. The engines roar.

Types of Sentences

When you write or talk, you use different kinds of sentences.

A **statement** tells something. It ends with a period.	*My father likes to cook.*
A **question** asks something. It ends with a question mark.	*What are we having for lunch?*
A **command** tells you to do something and ends in a period.	*Help me set the table.*
An **exclamation** shows strong feeling. It ends with an exclamation mark.	*This will be the best meal ever!*

Your Turn **Write each sentence. Tell what type it is.**

1. Can my friend join us for dinner?
2. Pasta is his favorite meal.
3. I can't believe how hungry we are!
4. Call us when it's ready.

Subjects in Sentences

Every sentence has two parts. The **complete subject** in a sentence tells what or whom the sentence is about. The complete subject of a sentence can be more than one word.

Rain *pours down from the sky.*
All of my friends *run for cover.*

Your Turn **Write each sentence. Draw a line under the complete subject of each sentence.**

1. Lightning flashes.
2. Thunder rumbles in the distance.
3. A gust of wind shakes the tree.
4. Sheets of rain strike the window.
5. Do storms like these happen often?

Predicates in Sentences

Every sentence has a subject and a **predicate**. The **complete predicate** tells what the subject does or is. The complete predicate can be one or more words.

The dog **howled**.
The girl **will feed him dinner**.

Your Turn **Write each sentence. Draw a line under the complete predicate of each sentence.**

1. The cat sneezes.
2. The bird flaps its wings.
3. The thirsty dog pushes his bowl.
4. My friends and I laugh.
5. Did the dog smile?

Sentences

Simple Sentences

A simple sentence expresses only one idea. It contains only one subject and one predicate.

The young man wants to fly airplanes.

Your Turn **Write each sentence. Circle the subject and underline the predicate.**

1. He reads about airplanes.
2. His friend was a pilot.
3. The man enrolls in a training program.
4. All of his classmates share his interest.
5. It will be quite an adventure!

Compound Sentences

A **compound sentence** expresses two or more complete thoughts, which are **combined** using a comma and a **coordinating conjunction**. A coordinating conjunction is a small word that links two things, such as *and, but,* or *or.*

I walk to school. You ride the bus. (two simple sentences)
I walk to school, **but** *you ride the bus. (one compound sentence)*

Your Turn **Write each pair of simple sentences as a compound sentence.**

1. Sam runs down the hall. The dog chases him.
2. Tara goes outside. It starts to rain.
3. Tara gets a ride. Sam decides to walk in the rain.
4. The school bell rings. The students go inside.
5. The rain turns to snow. Students are sent home early.

Complex Sentences

A **complex sentence** has an independent clause and one or more dependent clauses. An **independent clause** contains a subject and predicate and tells a complete thought.
A **dependent clause** contains a subject and predicate, and begins with a **subordinating conjunction**, such as *after, although, before, because, if, since, until, when, where, while.* Place a **comma** after the dependent clause when it is at the start of a sentence. *When we got there, the game began.*

Your Turn Write each sentence. Circle the independent clause. Underline the dependent clause.

1. People stopped to talk when they saw us there.
2. Because we sold so many, we raised a lot of money.

Run-On Sentences

A **run-on sentence** contains two or more independent clauses without the proper conjunctions or punctuation.
I cooked dinner my family liked it they asked for seconds.
You can correct run-on sentences using these strategies.

Break the independent clauses into separate sentences.	*I cooked dinner. My family liked it. They asked for seconds.*
Create a compound subject or compound predicate.	*I cooked dinner. My family liked it and asked for seconds.*
Create a compound sentence using coordinating conjunctions.	*I cooked dinner, and my family liked it. They asked for seconds.*
Create a complex sentence using subordinating conjunctions.	*I cooked dinner. Since my family liked it, they asked for seconds.*

Your Turn Correct each run-on sentence.

1. This was my first time cooking my mother helped.
2. Did you make the dessert I didn't have time.
3. I cooked the meal I didn't have to do the dishes.

Nouns

Common and Proper Nouns

A **noun** names a person, place, thing, or idea. A **common noun** names any person, place, or thing. A **proper noun** names a particular person, place, or thing, and begins with a capital letter.

The **student** studied the **atlas**. (common)
Tom located **Mount Rushmore**. (proper)

Your Turn Write each sentence. Underline each common noun. Circle each proper noun.

1. The book had many maps.
2. Can Joanne find our state?
3. Which country in Africa did your parents visit?
4. The library has a huge globe.

Concrete and Abstract Nouns

A **concrete noun** names a person, place, or thing that can be seen or touched. An **abstract noun** names a quality, concept, or idea that cannot be seen or touched. Many abstract nouns have no plural form.

Barbara set the **book** on the **desk**. (concrete)
Reading gives me great **joy**. (abstract)

Your Turn Write each sentence. Underline each concrete noun. Circle each abstract noun.

1. My cousin is afraid of the dark.
2. The dog slept under his bed.
3. His father read for an hour.
4. Silence filled the house.

Singular and Plural Nouns

A **singular noun** names one person, place, thing, or idea. A **plural noun** names more than one. Add *-s* to form the plural of most nouns. Add *-es* to form the plural of nouns ending in *s, x, ch,* or *sh.* If a noun ends in a consonant + *y,* change *y* to *i* and add *-es.* If a noun ends in a vowel + *y,* add *-s.* If a noun ends in *-f,* you sometimes change *f* to *v* and add *-es.*

girl girls	*store stores*	*bench benches*
city cities	*key keys*	*leaf leaves*

Your Turn Write each sentence. Underline each singular noun. Circle each plural noun.

1. My mother bought two jackets.
2. Look at those hats near the door.
3. The boys found bats and balls.

Special Plurals and Collective Nouns

Some nouns change their spelling to name more than one. Others don't change at all.

man men	*woman women*	*child children*
tooth teeth	*mouse mice*	*foot feet*
fish fish	*sheep sheep*	*goose geese*
moose moose	*deer deer*	

A **collective noun** names a group that acts together as one thing. It can be singular or plural.

*Our **band** will play after the three other **bands**.*

Your Turn Write each sentence. Use the correct form of the noun in ( ).

1. As a joke, we played "Three Blind (Mouse)" first.
2. All the (man) and (woman) laughed.
3. (Child) danced in front of the stage.

Nouns

Singular and Plural Possessive Nouns

A **possessive noun** is a noun that shows who or what owns or has something. Add an **apostrophe (')** + s to a singular noun to make it possessive.

*My **dentist's** office is in my **uncle's** town.*

Add an **apostrophe** to the end of most plural nouns to make them possessive. Add an **apostrophe** + s to form the plural of possessive nouns that don't end in -s.

*The **children's** mother visited three **teachers'** rooms.*

Your Turn **Write each sentence. Change the word in ( ) into a possessive noun.**

1. I like listening to my (parents) stories.
2. They talk about my (family) adventures.
3. They told me about my (aunt) wedding.
4. My mother sat at the (children) table.

Combining Sentences: Nouns

Two sentences can be combined by joining two nouns in the subject or predicate with the conjunction *and*.

*Kendra went to her room. Scott went to his room. Kendra **and** Scott went to their rooms.*

Your Turn **Combine the nouns in the sentence pairs to form one sentence.**

1. Dad went to the city. Mom went to the city.
2. Dad visited a museum. Dad visited a gallery.
3. They saw a movie. They saw a jazz concert.
4. Did Mom enjoy it? Did Dad enjoy it?

Verbs

Action Verbs

An **action verb** is a word that shows action. Some action verbs tell about actions that are hard to see.

*The boy **dreamed** that he **scored** the winning goal.*

Your Turn Write each sentence. Underline each action verb.

1. He practices hard every day.
2. The team runs across the field.
3. The coach carries a bag of soccer balls.
4. He drops one and kicks it hard.
5. Look at how quickly the team chases it!

Linking Verbs

A **linking verb** does not express action. A linking verb connects a subject to another part of the sentence. The most common present-tense linking verbs are *am, are,* and *is*. The most common past-tense linking verbs are *was* and *were*.

*Roberto **is** a writer. His last two books **were** wonderful.*

Your Turn Write each sentence and underline each linking verb.

1. Carla is a fast reader.
2. Her book shelves are full.
3. Both of her parents are teachers.
4. They were happy to loan us books.
5. Carla learned to read when she was four years old.

Verbs

Present-Tense Verbs

Present-tense verbs tell what is happening now. When the subject is not *I* or *you*, add -*s* to tell what one person or thing is doing. Add -*es* to verbs that end in *sh, ch, s, z,* or *x*. Change *y* to *i* and add -*es* to verbs that end with a consonant and *y*.
   Jen **looks** up and **watches** the birds. She **studies** them.

Your Turn **Write each sentence in the present tense. Use the correct form of the verb in ( ).**

1. My older sister (know) how to repair things.
2. She (fix) clocks and watches.
3. My sister (try) to figure out what's wrong.
4. She (make) a list and (cross) out each item.

Past-Tense Verbs

Past-tense verbs tell about an action that has already happened. Add -*ed* to most verbs to form the past tense.
   The farmer **planted** the seeds last week.

If the verb ends in a consonant plus *y*, change the *y* to *i* before adding -*ed*. If the verb ends with a vowel and a consonant, double the final consonant before adding -*ed*. If the verb ends with a consonant and *e*, drop the *e* before adding -*ed*.
   She **raised** her cup and **sipped** from the cup she **carried**.

Your Turn **Write each sentence in the past tense. Use the correct form of the verb in ( ).**

1. We (hurry) to finish our work.
2. She (want) to be done before dark.
3. The farmer (tug) at the weeds.
4. She and I (care) for the plants every day.

Future-Tense Verbs

A **future-tense verb** tells about action that is going to happen. Use the verb *will* to write about the future.

*My family **will visit** the park this weekend.*

Your Turn Write each sentence in the future tense. Use the correct form of the verb in ( ).

1. Next summer I (take) my first train ride.
2. My family (cross) the country by car.
3. We (stay) with my grandparents.
4. My sister and I (learn) how to sail.
5. It (be) the best summer vacation ever!

Subject-Verb Agreement

A **present-tense verb** must agree with its subject. Do not add -*s* or -*es* to a present-tense verb when the subject is plural or *I* or *you*.

*My **sister sends** me a message, and **I answer** right away.*

Your Turn Write each sentence in the present tense. Use the correct form of each verb in ( ).

1. She (write) e-mails quickly.
2. My brother (spend) all day on the phone.
3. My mother (wish) he would clean his room.
4. "I'll do it later," he (say).
5. My sister (advise) him to do it now.

Verbs

Verbs *Be*, *Do*, and *Have*

The verbs *be, do,* and *have* have special forms.

Subject	Present Tense	Past Tense
I	*am, do, have*	*was, did, had*
You	*are, do, have*	*were, did, had*
Singular subjects	*is, does, has*	*was, did, had*
We	*are, do, have*	*were, did, had*
Plural subjects	*are, do, have*	*were, did, had*

I ***am*** *happy with my phone, but his phone* ***has*** *a problem.*
Were *the pages you* ***had*** *clear, or* ***did*** *you need new ones?*

Your Turn **Write each sentence. Use the correct form of the verb in ( ).**

1. You (be) beginning to feel sleepy now.
2. I (be) standing beside you.
3. Your brother (have) an alarm clock once.
4. Where (do) he keep it now?

Main and Helping Verbs

Sometimes a verb can be more than one word. The **main verb** tells what the subject is or does. The **helping verb** helps the main verb show an action. *Have* and *be* can be used as helping verbs.

Beth ***is writing*** *a letter. She* ***had sent*** *one already.*

Your Turn **Write each sentence. Circle the helping verb. Underline the main verb.**

1. We are going on a field trip.
2. The class will visit a television studio.
3. My cousin has worked there for years.

Contractions with *Not*

A **contraction** combines a verb with the word *not*. An **apostrophe (')** takes the place of the missing letters. Some contractions have irregular spellings, such as *won't* (will not) and *can't* (cannot).

*Carol **isn't** feeling well. She **hasn't** eaten all day.*

Your Turn **Write each sentence. Combine the two words in ( ) to form a contraction.**

1. She (did not) bring a lunch today.
2. I (do not) understand why.
3. This (is not) the first time.
4. She (does not) know if she should buy lunch.
5. (Will not) someone share a sandwich with her?

Combining Sentences: Verbs

When two sentences have the same subject, you can combine the predicates. Use the conjunction **and** to join the predicates.

***Joe** runs to the door. **Joe** opens it.*
*Joe runs to the door **and** opens it.*

Your Turn **Combine the predicates in the sentence pairs to form one sentence.**

1. The boy stumbled. The boy fell.
2. The teacher raced over. The teacher helped him up.
3. He brushed off his clothes. He laughed.
4. The boy went to the bench. He sat down.
5. We went over. We made sure he was fine.

Verbs

Irregular Verbs

An **irregular verb** has a special spelling to show the past tense. Some verbs have a special spelling when used with the helping verb *have*.

Present	Past	Past with the verb *have*
am/are/is (be)	was/were	been
begin	began	begun
bring	brought	brought
come	came	come
do/does (do)	did	done
eat	ate	eaten
give	gave	given
go	went	gone
grow	grew	grown
hide	hid	hidden
run	ran	run
say	said	said
see	saw	seen
sing	sang	sung
sit	sat	sat
tell	told	told

Your Turn **Write each sentence. Use the correct form of the verb in ( ).**

1. Last night, our show (begin) late.
2. The audience (sit) patiently in their seats.
3. My parents had (come) to see the show.
4. I (tell) them to sit near the front.
5. They have (see) me in many plays before.

Pronouns

Pronouns

Pronouns take the place of nouns. A **personal pronoun** refers to specific people or things. An **indefinite pronoun** does not refer to specific people or things.

Personal pronouns: *I, you, he, she, it, we, they, me, her, him, us, them*

Common indefinite pronouns: *anyone, anything, everyone, no one, nothing, somebody, something*

> *We promised to tell no one. I won't say anything about it.*

Your Turn Write each sentence. Circle the personal pronouns. Underline the indefinite pronouns.

1. We talked to the police officer.
2. Something didn't seem right to us.
3. She gave me a phone number to call.
4. I could talk to anyone at the station.
5. "They want to help everyone," she told us.

Singular and Plural Pronouns

Singular pronouns include *I, you, he, she, it, me, her,* and *him.* Indefinite pronouns are usually treated as singular.

Plural pronouns include *we, you, they, us,* and *them.*

> *He* gives *them* money. *They* give *him* change. *Everybody is happy.*

Your Turn Write each sentence. Circle the singular pronouns. Underline the plural pronouns.

1. We wanted to give my mother a gift.
2. I found it on sale at the mall.
3. I bought two because they were on sale.
4. We hope that she likes them.

Pronouns

Subject Pronouns

A **subject pronoun** tells who or what did an action in a sentence. *I, you, he, she, it, we,* and *they* are subject pronouns.
She *calls us into the dining room.* **We** *are ready to eat.*

Your Turn Write each sentence. Change the words in ( ) into subject pronouns.

1. (My grandmother) cooks our dinner.
2. (My brother and sister) sit at the table.
3. Can (my brother) have a glass of juice?
4. (My sister and I) have a surprise dessert for our parents.
5. No one enjoys food as much as (my father) does!

Object Pronouns

An **object pronoun** comes after an action verb or prepositions such as *for, at, of, on, with, about,* and *to.* The pronouns *me, you, him, her, it, us,* and *them* are object pronouns.
The clerk gave **us** *two tickets. My parents put* **them** *in a safe place.*

Your Turn Write each sentence. Change the words in ( ) into object pronouns.

1. My parents called (my sister and me) downstairs.
2. We gave (my parents) our attention.
3. They looked at (my sister).
4. "Tell (the two children) about our vacation," Mom said.
5. "Let's start (our vacation) today!" Dad exclaimed.

Pronoun-Verb Agreement

A present-tense action verb must **agree** with its subject pronoun. Add -s to most action verbs in the present tense when you use the pronouns *he, she,* and *it.* Do not add -s to an action verb in the present tense when you use the pronouns *I, we, you,* and *they.*

He understands *what* **they say.**

Your Turn **Write each sentence. Use the correct present-tense forms of the verbs in ( ).**

1. She (create) a secret code.
2. We (help) write a message.
3. It (give) directions to a special place.
4. Did you (try) to figure it out?
5. They (succeed) and (find) the hidden treasure.

Possessive Pronouns

A **possessive pronoun** takes the place of a possessive noun. It shows who or what owns something. *My, your, her, his, its, our,* and *their* are possessive pronouns that are used before other nouns. *Mine, yours, his, hers, its, ours,* and *theirs* are possessive pronouns that can stand alone.

Jim used **his** *camera on* **our** *trip.* **Mine** *was in* **my** *room.*

Your Turn **Write each sentence. Change the words in ( ) into possessive pronouns.**

1. I borrowed (my sister's) camera.
2. She showed me how to use (the camera's) buttons.
3. I saw three hawks and took (the hawks') picture.
4. Mine came out better than (your pictures).
5. Shall we show (your and my) pictures to Mom?

Pronouns

Pronoun-Verb Contractions

A **contraction** is a shortened form of two words that are combined. An **apostrophe** (') replaces missing letters.
It's hard work, but you'll be proud when we're done.

Your Turn **Write each sentence. Form contractions.**

1. (I will) be working at the animal shelter today.
2. (They are) having a dog wash.

Reflexive Pronouns

A **reflexive pronoun** shows that a subject does something for or to itself. The ending *-self* is used for singular pronouns. The ending *-selves* is used for plural pronouns.
*Mom made **herself** a snack. We got **ourselves** one, too.*

Your Turn **Write each sentence. Replace the words in ( ).**

1. My father teaches (my father) sign language.
2. We like to give (to us) new challenges.
3. I found (for me) a tutor.

Pronoun-Antecedent Agreement

The **antecedent** is the noun or nouns to which a pronoun refers. A pronoun must agree with the number and gender of its antecedent.
My sister thinks she is late. My brothers think they are late.

Your Turn **Write each sentence. Choose the correct pronoun. Then underline the antecedent.**

1. My father says that (he, she) will go jogging.
2. My friend and I decide that (he, we) will join him.

Adjectives

Adjectives

An **adjective** is a word that describes a noun. Some adjectives tell **what kind** or **how many**. *Few, many,* and *several* are special adjectives that tell how many.

> **Several** students with **blue** banners ran onto the **crowded** field.

Your Turn **Write each sentence. Circle each adjective and underline the noun being described.**

1. It was an exciting game.
2. The band played a happy song.
3. We gave the team a huge cheer.
4. Three players lifted up the coach.
5. Did you see who scored the final goal?

Articles

The words **the, a,** and **an** are special adjectives called **articles**. Use *a* before singular nouns that begin with a consonant. Use *an* before singular words that begin with a vowel. Use *the* before singular and plural nouns.

> **An** otter left **a** trail along **the** river.

Your Turn **Write each sentence. Circle each article.**

1. We followed the animal's tracks.
2. Dad was sure it was a deer.
3. We looked through an opening in the bushes.
4. A baby moose stood in the meadow.
5. I took a picture of it for the school paper.

Adjectives

This, That, These, and Those

This, that, these, and *those* are special adjectives that tell **how many** and **how close**. ***This*** and ***that*** refer to singular nouns. ***These*** and ***those*** refer to plural nouns.

*I drew **this** sketch here and **those** sketches over there.*

Your Turn **Write each sentence. Choose the correct adjective in ( ) to complete the sentence.**

1. My mother likes (this, these) painting.
2. My father prefers (that, those) posters.
3. (This, These) rooms need more art.

Adjectives That Compare

Add *-**er*** to an adjective to compare two nouns. Add *-**est*** to compare more than two nouns. When the adjective ends in a consonant and *y*, change the *y* to *i* before adding *-er* or *-est*. When the adjective ends in *e*, drop the *e* before adding *-er* or *-est*. For adjectives that have a single vowel before a final consonant, double the final consonant before adding *-er* or *-est*. For long adjectives, use ***more*** in front of the adjective instead of adding *-er*. Use ***most*** in front of the adjective instead of adding *-est*.

*I chose the **smallest** and **scariest** book on the list.*
*My sister picked a **more difficult** book than mine.*

Your Turn **Write each sentence. Use the correct form of the adjective in ( ).**

1. I wrote a (short) book report than my sister's.
2. I asked the (smart) student in our class for help.
3. He had chosen the (easy) book of all.
4. This book is (interesting) than the first one.

Adverbs

Adverbs

An **adverb** is a word that tells more about a verb. Adverbs tell *how*, *when*, or *where* an action takes place.

> She **often** reads **quietly** in class.

Your Turn Write each sentence. Circle each adverb. Then underline the verb it tells about.

1. The boat moved slowly across the sea.
2. Sea gulls flew nearby.
3. A whale suddenly swam to the surface.
4. We ran fast to take a picture.
5. Soon it disappeared underwater.

Adverbs That Tell *How*

Some adverbs tell **how** an action takes place. Adverbs that tell *how* often end with *-ly*.

> The telephone rang **loudly**.

Your Turn Write each sentence. Circle each adverb. Then underline the verb it tells about.

1. I quickly answered the phone.
2. A man talked rapidly to me.
3. I tried hard to understand.
4. I calmly asked him to repeat himself.
5. I wrote the information carefully.

Adverbs

Adverbs That Tell *When* or *Where*

Some adverbs tell *when* or *where* an action takes place.
Later, we went *outside* for a walk.

Your Turn **Write each sentence. Circle each adverb. Then underline the verb it tells about.**

1. I soon heard a sound.
2. Then I looked up.
3. A plane appeared overhead.
4. Next I saw parachutes open.
5. Sky divers eventually landed nearby.

Adverbs That Compare

For most adverbs, use *more* in front of the adverb to compare two actions. Use *most* in front of the adverb to compare more than two actions. For shorter adverbs, add *-er* to the adverb to compare two actions. Add *-est* to compare more than two actions. When the adverb ends in a consonant and *y*, change the y to *i* before adding *-er* or *-est*.
*I ran **faster** than my friend, but he swam **more gracefully**.*

Your Turn **Write each sentence. Use the correct form of the adverb in ( ).**

1. I practiced (hard) than my brother did.
2. My father may arrive (soon) than my mother.
3. My mother drives (carefully) than he does.
4. I speak (quietly) in the car than I do on the field.
5. Dad listens (closely) of all the parents I know.

Prepositions

Prepositions

A **preposition** comes before a noun or a pronoun. A preposition shows how the noun or pronoun is linked to another word in the sentence. Some common prepositions are *in, into, at, of, from, on, with, to,* and *by*.

The girl **in** the raincoat ran **to** the house.

Your Turn **Write each sentence. Circle each preposition.**

1. Clouds appeared in the sky.
2. Rain fell on the town.
3. We were at the cinema.
4. We called our parents from the lobby.
5. They would meet us by the front door.

Prepositional Phrases

A **prepositional phrase** is a group of words that begins with a preposition and ends with a noun or pronoun. The noun or pronoun is the **object** of the preposition. A prepositional phrase can be used as an adjective or an adverb.

The man **at the beach** (adjective) swam **in the water** (adverb).

Your Turn **Write each sentence. Underline each prepositional phrase. Then circle the object of the preposition.**

1. Fish jumped into the air.
2. Children played with their friends.
3. A girl in sandals gathered shells.
4. A lifeguard watched us from his chair.
5. You found a starfish and showed it to me.

Mechanics: Abbreviations

Titles

An **abbreviation** is the shortened form of a word. It usually begins with a capital letter and ends with a period. You can abbreviate the titles before a name.

Mr. *Scott Fletcher*   **Ms.** *Bethany Collins*
Mrs. *Miko Anaki*   **Dr.** *Steinberg*

Your Turn **Write each name using the correct abbreviation and capitalization.**

1. mrs Patricia Barry
2. Mister Ed West
3. Doctor Paula hanson
4. ms susie blair

Days of the Week

You can abbreviate the days of the week.

Sun.   Mon.   Tues.   Wed.   Thurs.   Fri.   Sat.

Months of the Year

You can abbreviate the months of the year. Do not abbreviate the months May, June, or July.

Jan. Feb. Mar. Apr. Aug. Sept. Oct. Nov. Dec.

Your Turn **Write each sentence with the correct abbreviation.**

1. Our first game was on September 19.
2. We go on Monday and Tuesday this week.
3. Can you be here on January 22?
4. I can't make a Wednesday meeting.
5. We will finish on Thursday, March 15.

States

When you write an address, you may use United States Postal Service abbreviations for the names of states. The abbreviations are two capital letters with no period at the end. Do not use a comma between the city and the state when using these abbreviations.

Alabama	AL	Kentucky	KY	Ohio	OH
Alaska	AK	Louisiana	LA	Oklahoma	OK
Arizona	AZ	Maine	ME	Oregon	OR
Arkansas	AR	Maryland	MD	Pennsylvania	PA
California	CA	Massachusetts	MA	Rhode Island	RI
Colorado	CO	Michigan	MI	South Carolina	SC
Connecticut	CT	Minnesota	MN	South Dakota	SD
Delaware	DE	Mississippi	MS	Tennessee	TN
District of		Missouri	MO	Texas	TX
Columbia	DC	Montana	MT	Utah	UT
Florida	FL	Nebraska	NE	Vermont	VT
Georgia	GA	Nevada	NV	Virginia	VA
Hawaii	HI	New Hampshire	NH	Washington	WA
Idaho	ID	New Jersey	NJ	West Virginia	WV
Illinois	IL	New Mexico	NM	Wisconsin	WI
Indiana	IN	New York	NY	Wyoming	WY
Iowa	IA	North Carolina	NC		
Kansas	KS	North Dakota	ND		

Your Turn Write the U.S. Postal Service Abbreviation for each of the following.

1. Sacramento, California
2. Ft. Worth, Texas
3. Sarasota, Florida
4. Springfield, New Jersey
5. Portland, Maine

Mechanics: Capitalization

First Word in a Sentence

Capitalize the first word in a sentence. Capitalize the first word of a direct quotation, or the exact words a person says.
Maura said, "This is my new backpack."

Your Turn **Write each sentence. Use capital letters.**

1. the dog chewed through her old backpack.
2. she said, "he didn't mean to do it."
3. "was he hungry?" I asked.
4. This one is nice, But the old one was better.

Letters

Capitalize all of the major words in a letter's greeting. Capitalize only the first word in the closing of a letter.
Dear Sir or Madam: Sincerely yours,

Proper Nouns: Names and Titles of People

Capitalize names and initials that stand for names. Capitalize titles or abbreviations of titles when they come before the names of people. Capitalize the pronoun *I*.
Mr. Renaldo and I paid Dr. E. M. Carlson a visit.

Your Turn **Write each part of the letter with the correct capitalization.**

1. dear mrs. price,
2. t. j. and i are doing a survey.
3. Have you read gov. grant's new book?
4. very truly yours,

Proper Nouns: Names of Places

Capitalize the names of cities, states, countries, and continents. Do not capitalize articles or prepositions that are part of the names.

We flew from the **United States of America** to **Africa**.

Capitalize the names of geographical features.

Lake Superior is north of the other **Great Lakes**.

Capitalize the names of streets and highways. Capitalize the names of buildings and bridges.

Golden Gate Bridge   **Empire State Building**

Capitalize the names of stars and planets.

I asked, "How far from **Earth** is the star **Sirius**?"

Your Turn **Rewrite each sentence. Use capital letters where needed.**

1. Our teacher used to live in colorado.
2. Lucy once visited the grand canyon.
3. A telescope can help you see rigel, a bright star.
4. The observatory is near la plata river.
5. The night sky in north america is different from australia.

Mechanics: Capitalization

Other Proper Nouns

Capitalize the names of schools, clubs, teams, businesses, and products.

   *I joined the **Health Club** at **Parker Elementary**.*

Capitalize the days of the week, months of the year, and holidays. Do not capitalize the names of the seasons.

   ***Labor Day** is the first **Monday** in **September**.*

Capitalize abbreviations when they refer to people or places.

   ***Ms. Webb** studied the **Mt. St.** Helens eruption.*

Capitalize the first, the last, and all important words in the title of a book, poem, article, song, short story, film, video, television program, and newspaper.

   *She wrote an article called "**A Year** in the **Danger Zone**."*

Your Turn **Rewrite each sentence. Use capital letters where needed.**

1. We saw the film "volcano!" on monday.
2. The young geologists club hosted.
3. After the movie, mr. Isham gave a talk.
4. He will come back after memorial day.
5. Can we go to elmwood middle school to hear him?

Mechanics: Punctuation

End Marks for Sentences

A **period (.)** ends a statement or a command.
A **question mark (?)** ends a question.
An **exclamation mark (!)** ends an exclamation.
> *Why is that light flashing?* *I'm scared!* *It was a false alarm*.

Your Turn **Write each sentence. Use the correct end punctuation.**

1. We heard the siren outside
2. It was incredibly loud
3. Who called the fire department
4. Do they know it's a false alarm
5. Go ask the teacher for advice

Periods for Abbreviations

Use a period at the end of an abbreviation. Use a period after initials.
> *We will visit* **Dr. E. X.** *Young on* **Oct.** *21.*

Your Turn **Write each sentence. Insert periods where needed.**

1. Ms Mehta teaches us how to cook.
2. We'll return on Apr 13 for our next class.
3. Prof Ruiz is also in our class.
4. We meet both Mon and Tues next week.
5. My brother A J will drive to Pine St to get us.

Mechanics: Punctuation

Colons

Use a **colon** after the greeting of a business letter.
Dear Senator:

Use a **colon** to separate the hour and minute when you write the time of day.
Thank you for speaking with me at 10:30 today.

Your Turn **Write each sentence. Insert colons.**

1. Dear Judge Michaels
2. I'm sorry I missed our 930 interview.
3. The 830 bus was late this morning.

Commas in Letters

Use a **comma (,)** between the name of a city and the complete name of a state. Do not use a comma before the postal service abbreviation for a state.
Plymouth, Massachusetts   Tucson AZ

Use a **comma** between the day and the year in a date.
March 6, 2005   Tuesday, June 12, 2010

Use a **comma** after the greeting and closing in a friendly letter.
Dear Uncle Walter,   Sincerely yours,

Your Turn **Write each sentence. Add commas where needed.**

1. Dear Aunt Jobeth
2. Have you ever been to Miami Florida?
3. We moved there on July 15 2011.
4. We had been living in Fort Worth Texas.

Commas in Sentences

Use a **comma** before *and, but,* or *or* in a compound sentence.
Use a comma after a dependent clause at the start of a sentence.

> *When the bell rang, the students looked up, and the teacher spoke.*

Use **commas** to separate three or more items in a series.
> *We put away the toys, games, and craft materials.*

Use a **comma** after the words *yes* or *no* when they begin a sentence. Use a comma after the name of a person being spoken to.
> ***Yes,*** *my desk is clean.*   ***Adam****, is your desk clean?*

Your Turn **Write each sentence. Add commas.**

1. I used yellow red and blue for my poster.
2. Jen started a new one and I finished it.
3. No it wasn't as good as the first one.
4. Is there enough or should I make more?
5. Leroy I think we have more than enough!

Apostrophes

Use an **apostrophe (')** with a noun to show possession.
Use an apostrophe in a **contraction** to show where a letter or letters are missing.

> *Our uncle's horses aren't where they should be.*

Your Turn **Write each sentence. Add apostrophes.**

1. I got the keys to my mothers car.
2. We cant see them from the road.
3. "Whats causing that dust cloud?" I asked.
4. "Thats probably them," my mother said.
5. We could see all the horses tracks in the dirt.

Mechanics: Punctuation

Commas and Quotation Marks

Use **quotation marks** at the beginning and at the end of the exact words a person says. Use **commas** to set off a direct quotation.

"I like it," she said, "but it's too expensive."

Your Turn **Write each sentence. Add commas and quotation marks where necessary.**

1. I asked Aren't they on sale?
2. She replied I wish they were.
3. Let's try another store I said.
4. What about that one? she asked.
5. The price here she said is better!

Italics or Underlining

Use italics or underlining for the title of a book, movie, play, magazine, or newspaper.

I read <u>The Big Book of Adventures</u> at the library.

Your Turn **Write each sentence. Underline the titles.**

1. I was looking for The Little Book of Secrets.
2. I read a review of it in the Chicago Tribune.
3. Does the movie Tiny Truths still mention the book?
4. My aunt saw Big Lies, Little Lies on stage in New York.
5. I'm writing a book called The Whole Story.